Annual Meeting of the American Electrophoresis Society 2013

(AES)

Topical Conference at the 2013 AIChE Annual Meeting: Global Challenges for Engineering a Sustainable Future

San Francisco, California, USA
3-8 November 2013

ISBN: 978-1-63439-023-1

Printed from e-media with permission by:

Curran Associates, Inc.
57 Morehouse Lane
Red Hook, NY 12571

Some format issues inherent in the e-media version may also appear in this print version.

Printed by Curran Associates, Inc. (2013)

AIChE
3 Park Avenue
New York, NY 10016-5991

Phone: (203) 702-7660
Fax: (203) 775-5177

www.aiche.org

Additional copies of this publication are available from:

Curran Associates, Inc.
57 Morehouse Lane
Red Hook, NY 12571 USA
Phone: 845-758-0400
Fax: 845-758-2634
Email: curran@proceedings.com
Web: www.proceedings.com

TABLE OF CONTENTS

(34a) Identification of Activated Receptor Tyrosine Kinases Using 2DE Western Blot Image Overlays

Kendrick, N., Kendrick Labs Inc
Hoelter, M., Kendrick Labs Inc
Johansen, J., Kendrick Labs Inc
Originally presented on: 11/4/2013 8:30:00 - 8:55:00

Receptor tyrosine kinases (RTK) are a class of trans-membrane protein activated by phosphorylation of tyrosine groups on the cytosolic chain domain. Activated RTKs, often growth factors like EGFR (epidermal growth factor receptor), are known drivers of cancer growth. Several RTK inhibitors have been tested in clinical trials and are approved by the FDA for cancer treatment. However, physicians have difficulty matching patients to the correct RTK inhibitor during the critical period of early treatment. Cancer growth triggered by one kind of RTK does not respond to an inhibitor of a different kind. Other oncogenic proteins, RAS for example, trigger cancer growth by different pathways and also do not respond to RTK inhibitors. Tests for inactive RTKs, immunohistochemistry for example, do not correlate with inhibitor success. A test for detection and identification of specific, *activated* RTKs in tumor tissue is needed.

We have previously shown that tyrosine phosphorylation (activation) of putative RTKs can be detected in human tumors using carrier ampholine 2-dimensional gel electrophoresis (CA-2DE) in conjunction with Western blotting (WB). Initial attempts using mass spectrometry to identify activated RTKs detected by WB were unsuccessful. Attempts to identify the activated RTKs independently using protein arrays were also unsuccessful. RTKs are high molecular weight, membrane proteins that do not dissolve without the solubilizing agent, sodium dodecyl sulfate (SDS). Protein arrays are incompatible with SDS; CA-2DE is compatible with this detergent. The very high sensitivity of the WB method is due to the high affinity of the PY20 antibody for the phosphotyrosine epitope.

The approach taken in this work is to identify the activated RTKs by duplex 2D WB. First, CA-2DE WB with the anti-phosphotyrosine antibody is performed and images obtained. The blot is stripped, and then reprobed with an antibody against a known RTK such as EGFR. Because both WB images are obtained from the same membrane, they are superimposable and can be easily overlaid using software called SameSpots. Exact overlay of the phosphotyrosine WB (activation pattern) with the anti-RTK protein WB (identity pattern) confirms identification of the activated RTK. Mismatches are easy to detect. Successful results from analysis of a set of six human lung cancer tumors obtained from a tissue bank are discussed.

***** Formatting Issue - Best Available Paper/Graphic *****

(34b) On-Chip Protein Extraction and Albumin Exclusion From Plasma and Serum Using Isotachophoresis

Qu, Y., Stanford University
Marshall, L. A., Stanford University
Santiago, J. G., Stanford University
Originally presented on: 11/4/2013 8:55:00 - 9:20:00

harp counter migrating bands. (Bottomng by ion by using

ON-CHIP PROTEIN EXTRACTION AND ALBUMIN EXCLUSION FROM PLASMA AND SERUM USI

Yatian Qu, [1] Lewis A. Marshall, [2] Juan G. Santiago[1]

[1] Department of Mechanical Engineering, Stanford University

[2] Department of Chemical Engineering, Stanford University

Stanford, CA 94305, USA

Human blood plasma and serum contain thousands of proteins, which makes them ideal matrices for diagnostics an
removal of albumin from serum</title><secondary-title>Proteomics</secondary-title><alt-title>Proteomics</alt-tit
Fractionation</keyword><keyword>Chemical Precipitation</keyword><keyword>Chromatography, Affinity</key
urls><url>http://www.ncbi.nlm.nih.gov/pubmed/16145708</url></related-urls></urls><electronic-resource-num>
3C456E644E6F74653E3C436974653E3C417574686F723E4772616E6765723C2F417574686F723E3C59656172
F726569676E2D6B6579733E3C7265662D74797065206E616D653D224A6F75726E616C2041727469636C6522
F67792C20556E6976657273697479206F66204D6963686967616E2C2032323130204D65646963616C20536369
F7469746C653E3C7365636F6E646172792D7469746C653E50726F74656F6D6963733C2F7365636F6E646172
C2F616262722D313E3C2F616C742D706572696F646963616C3E3C70616765733E343731332D383C2F7061676
920416E616C797369732F2A6D574686F64733C2F6B6579776F72643E3C6B6579776F72643E50726F74656F6
0284C696E6B696E67293C2F6973626E3E3C616363657373696F6E2D6E756D3E31363238313138303C2F61636
86F723E4772616E6765723C2F417574686F723E3C596561723E323030353C2F596561723E3C5265634E756D3E
3C617574686F723E4772616E6765722C204A2E3C2F617574686F723E3C617574686F723E53696464697175692
2C20416E6E204172626F722C204D492034383130392D303630322C205553412E206A696C6C6740756D696368
746C653E3C2F7469746C65733E3C706572696F646963616C3E3C66756C6C2D7469746C653E50726F74656F6
265723E3C6B6579776F7264733E3C6B6579776F72643E4379746F6B696E65732F626C6F6F642F2A6368656D6
C2F6B6579776F72643E3C6B6579776F72643E536572756D20416C62756D696E2F2A6368656D69737472793C

*** Formatting Issue - Best Available Paper/Graphic ***

E687474703A2F2F7777772E6E6362692E6E6C6D2E6E69682E676F762F7075626D65642F31363238313138303
3C456E644E6F74653E3C436974653E3C417574686F723E4772616E6765723C2F417574686F723E3C59656172
F726569676E2D6B6579733E3C7265662D74797065206E616D653D224A6F75726E616C2041727469636C6522
F67792C20556E697665727369747920F66204D6963686967616E2C2032323130204D65646963616C205363696
F7469746C653E3C7365636F6E646172792D7469746C653E50726F74656F6D6963733C2F7365636F6E6461727
C2F616262722D313E3C2F616C742D706572696F646963616C3E3C70616765733E343731332D383C2F7061676
920416E616C797369732F2A6D6574686F64733C2F6B6579776F72643E3C6B6579776F72643E50726F74656F6
0284C696E6B696E67293C2F6973626E3E3C616363657373696F6E2D6E756D3E31363238313138303C2F61636
86F723E4772616E6765723C2F417574686F723E3C596561723E323030353C2F596561723E3C5265634E756D3E
3C617574686F723E4772616E6765722C204A2E3C2F617574686F723E3C617574686F723E5369646464697175692
2C20416E6E204172626F722C204D492034383130392D303630322C205553412E206A696C6C6740756D696368
746C653E3C2F7469746C65733E3C706572696F646963616C3E3C66756C6C2D7469746C653E50726F74656F6
265723E3C6B6579776F7264733E3C6B6579776F72643E4379746F6B696E65732F626C6F6F642F2A6368656D6
C2F6B6579776F72643E3C6B6579776F72643E536572756D20416C62756D696E2F2A6368656D69737472793C
E687474703A2F2F7777772E6E6362692E6E6C6D2E6E69682E676F762F7075626D65642F31363238313138303
3C456E644E6F74653E3C436974653E3C417574686F723E456368616E3C2F417574686F723E3C596561723E32
569676E2D6B6579733E3C7265662D74797065206E616D653D224A6F75726E616C2041727469636C65223E31
6464726573733E5468652057697374617220496E737469747574652C205068696C6164656C706869612C2050412
69746C653E50726F74656F6D6963733C2F616C742D7469746C653E3C2F7469746C65733E3C706572696F6469
3C2F766F6C756D653E3C6E756D6265723E31333C2F6E756D6265723E3C6B6579776F7264733E3C6B657977
6B6579776F72643E3C6B6579776F72643E48756D616E733C2F6B6579776F72643E3C6B6579776F72643E496
616C797369732F2A6D6574686F64733C2F6B6579776F72643E3C6B6579776F72643E50726F74656F6D653C2F
2F796561723E3C7075622D64617465733E3C646174653E4175673C2F646174653E3C2F7075622D64617465733
632D7265736F757263652D6E756D3E31302E313030322F706D69632E3230303430313232383C2F656C656374
3C456E644E6F74653E3C436974653E3C417574686F723E456368616E3C2F417574686F723E3C596561723E32
569676E2D6B6579733E3C7265662D74797065206E616D653D224A6F75726E616C2041727469636C65223E31
6464726573733E5468652057697374617220496E737469747574652C205068696C6164656C706869612C2050412
69746C653E50726F74656F6D6963733C2F616C742D7469746C653E3C2F7469746C65733E3C706572696F6469
3C2F766F6C756D653E3C6E756D6265723E31333C2F6E756D6265723E3C6B6579776F7264733E3C6B657977
6B6579776F72643E3C6B6579776F72643E48756D616E733C2F6B6579776F72643E3C6B6579776F72643E496
616C797369732F2A6D6574686F64733C2F6B6579776F72643E3C6B6579776F72643E50726F74656F6D653C2F
2F796561723E3C7075622D64617465733E3C646174653E4175673C2F646174653E3C2F7075622D64617465733
632D7265736F757263652D6E756D3E31302E313030322F706D69632E3230303430313232383C2F656C656374
T.</author><author>Wang, A. H.</author></authors></contributors><auth-address>Core Facilities for Proteomic
27</pages><volume>26</volume><number>11</number><keywords><keyword>Acetone</keyword><keyword>
Acid</keyword></keywords><dates><year>2005</year><pub-dates><date>Jun</date></pub-dates></dates><isb
protocol for protein extraction and exclusion of albumin from human plasma or serum samples. This method

We perform this protein extraction and albumin exclusion technique by using isotachophoresis (ITP). ITP achieves
9511</pages><volume>81</volume><number>22</number><dates><year>2009</year></dates><urls></urls></
remains unfocused in channel.

We have designed a large-volume microfluidic chip as shown in Figure 1 to demonstrate this technique. This chip i
additional buffering capacity.

We have successfully extracted proteins and eliminated albumin from 1 ÂµL plasma and serum samples by
Blue staining (Figure 3). We compared the extracted proteins to unprocessed plasma or serum proteins. We recover

*** Formatting Issue - Best Available Paper/Graphic ***

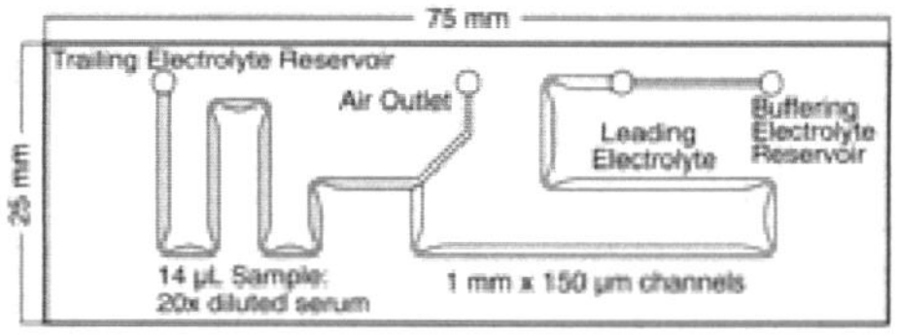

Figure 1. Microfluidics chip design for large-volume protein extraction and albumin exclusion from human plasma

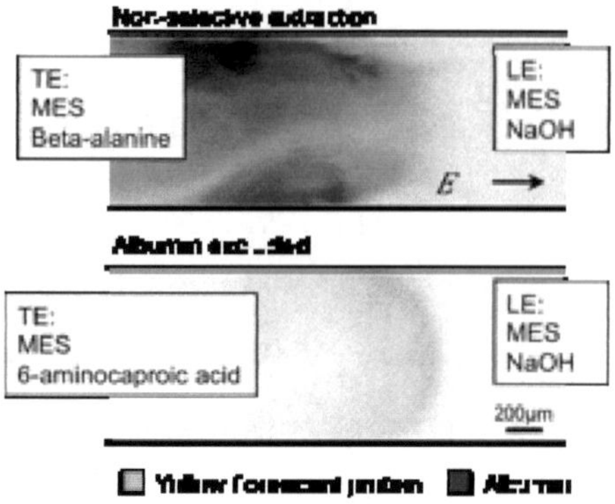

Figure 2. Comparison of non-selective and selective ITP protein extraction chemistries: (Top) Beta-alanine was us

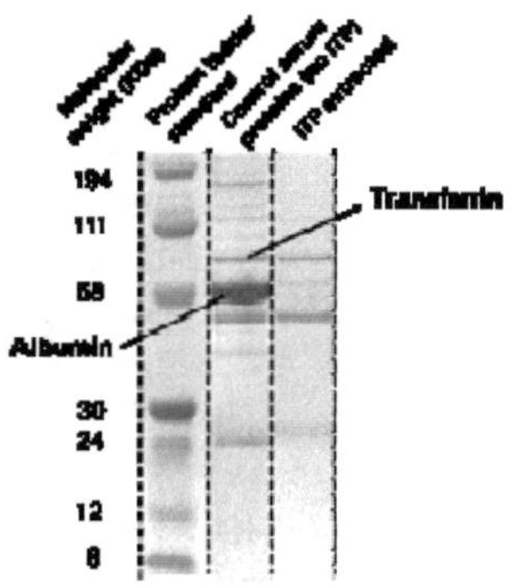

Figure 3. Off-chip SDS-PAGE separation images of proteins extracted with ITP and from control serum sa

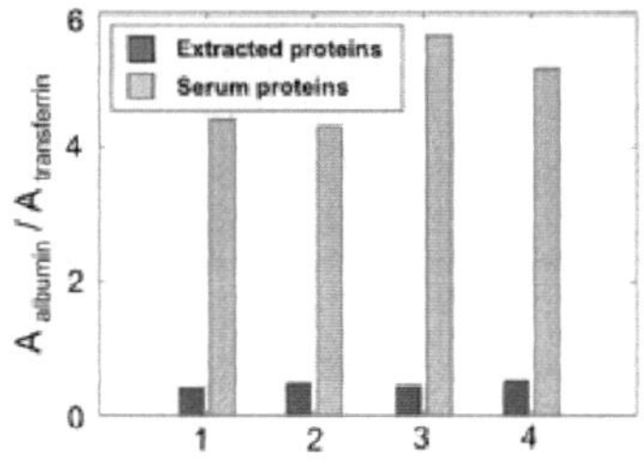

Figure 4. Protein absorbance analysis on SDS-PAGE gel. Shown are repetitions of albumin amount relative to trar

REFERENCES

ADDIN EN.REFLIST 1. Colantonio, D. A.; Dunkinson, C.; Bovenkamp, D. E.; Van Eyk, J. E., Effective removal

2. Granger, J.; Siddiqui, J.; Copeland, S.; Remick, D., Albumin depletion of human plasma also removes low abun

3. Echan, L. A.; Tang, H. Y.; Ali-Khan, N.; Lee, K.; Speicher, D. W., Depletion of multiple high-abundance protei

4. Chen, Y. Y.; Lin, S. Y.; Yeh, Y. Y.; Hsiao, H. H.; Wu, C. Y.; Chen, S. T.; Wang, A. H., A modified protein prec

5. Persat, A.; Marshall, L. A.; Santiago, J. G., PuriÃžcation of Nucleic Acids from Whole Blood Using Isotachoph

(34c) Microchannel Electrophoresis Analysis of Amyloid Protein Aggregation

Hestekin, C. N., University of Arkansas
Pryor, E., University of Arkansas
Moss, M. A., University of South Carolina
Originally presented on: 11/4/2013 9:20:00 - 9:45:00

Amyloids are involved in a variety of diseases including Alzheimer's Disease and diabetes. A number of technical challenges exist for the quantitative analysis of the different sizes of amyloid proteins which are present during the early stages of aggregation, therefore requiring new technologies to be developed for enhanced protein separations. Microchannel electrophoretic techniques have emerged as powerful tools for the quantitative analysis of proteins. In this study, we investigated the ability of capillary and microchip electrophoresis to monitor the early stages of amyloid aggregation using amyloid beta (Aβ) and amylin as a model amyloid-forming proteins. For the Aβ protein, sample preparation effected the size of the initial starting material as well and the rate of protein aggregation. In addition, electrophoretic conditions including the use of polymeric coating and separations matrices as well as aggregation conditions (agitation, salt, etc.) were determined to affect the rate of protein aggregation.

(34d) Fractionation of Nanocrystals for Femtosecond Nanocrystallography of Membrane Proteins

Ros, A., Arizona State University
Abdallah, B.,
Kupitz, C.,
Fromme, P.,
Originally presented on: 11/4/2013 9:45:00 - 10:10:00

Femtosecond nanocrystallography is an emerging technique with the potential to obtain structural information of membrane proteins without the need of growing large crystals. The latter represents the major challenge in membrane protein structure elucidation as membrane proteins are difficult to grow in large sizes suitable for conventional X-ray structure determination. However, due to the heterogeneity inherent to current crystallization techniques for membrane proteins, nanocrystallography experiments are hampered by broad size distributions obtained among smaller crystals. Here, we demonstrate a method based on insulator-based dielectrophoresis to isolate protein crystal size fractions on the order of 100 nm.

To drive this size-based separation, we exploit the influence of dielectrophoretic focusing in an electroosmotically driven microdevice. The device consists of an inlet channel into which a bulk crystal solution can be injected. An insulator constriction is positioned between the inlet and outlets to create a heterogeneous electric field evoking dielectrophoresis (DEP) as particles migrate through the microchannel. Upon entering the restriction region, particles are repelled from the channel walls due to negative DEP and focus in the center of the microdevice. Conversely, smaller particles experiencing weaker repulsion are prone to deflection into a series of outlet channels for effective sorting. The device itself is fabricated using poly(dimethylsiloxane) (PDMS) and standard soft lithography. DC potentials are applied to the microdevice reservoirs to induce electroosmotic flow (EOF) as well as electric field gradients at the constriction region for dielectrophoretic focusing.

This approach was applied to a solution of photosystem I (PSI) membrane protein crystals. PSI is a large membrane protein complex containing 36 proteins to which 381 cofactors (chlorophylls, carotenoids, quinones, and 4Fe4S clusters) are non-covalently attached. PSI crystals ranging in size from ~100 nm to 20 μm typically arise from a successful batch crystallization experiment. We used such crystallization products directly in the iDEP sorter and could demonstrate excellent sorting of nanometer-sized crystals into selected outlet channels. The sizes of the fractionated crystals were examined with dynamic light scattering (DLS) resulting in a radius of ~100 nm with high monodispersity. Furthermore, the crystallinity of the fractionated PSI nanocrystals was confirmed by Second Order Nonlinear Imaging of Chiral Crystals (SONICC).

(34e) Human Fluid Sample Pretreatment for Biomarker Discovery

Ross, P., Sonata Biosciences, Inc.
Jahnke, F., Sonata Biosciences, Inc.
Originally presented on: 11/4/2013 10:10:00 - 10:35:00

Human fluids – including plasma, serum, and cerebrospinal and pleural effusion fluids -- potentially are rich sources of protein biomarkers that can be used to indicate the presence or progression of disease or the response of disease to treatment regimens. Human fluids are enormously complex, with hundreds of thousands of different proteins, protein fragments and post-translational modifications that are present with concentrations that span ten orders of magnitude. Potential biomarkers are expected to be present at very low concentrations when compared with the abundant proteins, and the wide protein dynamic range far exceeds that of the analytical instrumentation used for the analysis, whether gel electrophoresis (such as DIGE) or bottom-up proteomic methods are used.

Nine protein biomarker panels have been introduced to date, but discovery of new ones has slowed considerably. We believe that one reason is that pretreatment methods applied to human fluid samples are inadequate. We review the extensive work described in the literature, and introduce new ones that we are developing at Sonata Biosciences. A particular emphasis is placed on treatment of the large volumes that are required to delve deeply into the proteome to uncover dilute protein biomarkers.

(34f) Using Gradient Insulator-Based Dielectrophoresis to Capture Small Molecular Weight Proteins

Yanashima, R., Arizona State University
Hayes, M. A., Arizona State University
Originally presented on: 11/4/2013 10:35:00 - 11:00:00

Dielectrophoresis (DEP) has the potential to serve as a separation and concentration technique for small volume samples. Dielectrophoresis has typically been applied to particles or cells, but recently analytes with small molecular weights such as proteins, have been targeted. Protein dielectrophoresis has been demonstrated, with the first observations of protein DEP trapping about twenty years ago by Washizu et. al. (*IEEE Trans. Ind. Appl.,* 1994, 30, 835-843). The smallest proteins to be captured thus far using techniques similar to ours are the proteins BSA at 66 kDa and streptavidin at 60 kDa. Proteins of smaller molecular weights have been studied, and even successfully manipulated using dielectrophoresis resulting in streaming (*Electrophoresis*, 2013, 34, 1085-1096), but discrete isolation in a pseudo steady state has not been shown. Here, we extend the range of capture down to 14.3 kDa by capturing lysozyme from chicken egg white, along with other small molecular weight proteins.

Dielectrophoresis is carried out on a microfluidic device consisting of an insulating sawtooth-patterned microchannel such that an inhomogeneous electric field is induced in the channel when a DC potential is applied across the device. The gradient of dielectrophoretic forces in our device arises from the varying distances of each successive gate within the device: as the gates become narrower, the ratio of the dielectrophoretic force to the electrophoretic force increases. When the dielectrophoretic force is great enough to counteract all other forces a protein experiences within the channel, immobilization and concentration occurs.

One protein of great interest in the medical field is Aβ (1-40) amyloids. Aggregates of Aβ amyloid and other amyloids have been implicated in numerous human diseases, including Alzheimer's disease. While it is known that the fully-developed Aβ amyloid fibrils are related to this disease, recent research has suggested the smaller oligomers and protofibrils are more prevalent in disease pathogenesis and are more cytotoxic. The proteins for our research were chosen based on their small molecular weights such that their capture conditions might be comparable to small oligomers of Aβ amyloid proteins. Future work includes studying the behavior of small Aβ amyloid oligomers in our device.

(102a) Fluidic Dielectrophoresis: Polarization, Manipulation and Biosensing At Electrical Liquid Interfaces

Gagnon, Z. R., Johns Hopkins University
Originally presented on: 11/4/2013 12:30:00 - 12:52:00

To date, AC electric fields have been exploited to dielectrophoretically manipulate bubbles, liquid drops, particles, biomolecules and cells in microfluidic devices. Research and applications in this area, however, has been primarily limited to the interfaces formed between two immiscible metal-liquid, particle-liquid, or gas-liquid surfaces on particles. The influence of AC electric fields across aqueous liquid-liquid interfaces remains relatively unexplored. As the majority of microfluidic applications involve aqueous liquid flows at low Reynolds number, fluid interfaces formed between co-flowing fluid streams are a natural occurrence in microfluidic devices. Fundamentally, many electrokinetic phenomena arise from discontinuities in ionic flux and charge accumulation at electrical interfaces. Here, we explore the influence of AC electric fields on the electrical interface created between two aqueous liquids with disparaging electrical properties.

Using a microfluidic channel with embedded electrodes, two fluid streams - one with a greater electrical conductivity, the other a greater dielectric constant - were made to flow side-by-side. An AC field was applied across the flow channel and fluid was observed to displace across the phase interface. The displacement direction is frequency dependent, and is attributed to the Maxwell-Wagner interfacial polarization at the liquid-liquid electrical interface. At low AC frequency (< 1 MHz), below the interfacial charge relaxation time, the high conductive stream is observed to displace into the high dielectric stream. Above this frequency, the direction of liquid injection reverses, and the high dielectric stream injects into the high conductivity stream. Hence, a liquid crossover frequency, very similar to a dielectrophoretic crossover frequency, is observed and is well described by classical Maxwell-Wagner polarization mechanics.

In this talk, we demonstrate the use of interfacial polarization at liquid-liquid electrical interfaces for fluidic manipulation and biosensing applications, and demonstrate how these small microfluidic force fields can be used to regulate on-chip solute concentration levels and measure bioaffinity reactions non-optically.

(102b) Continuous Particle Sorting Using Three Dimensional Insulator Based Dielectrophoresis

Wang, Q., Massachusetts Institute of Technology
Buie, C. R., Massachusetts Institute of Technology
Originally presented on: 11/4/2013 12:52:00 - 13:14:00

Low frequency insulator based dielectrophoresis (iDEP), the induced polarizable particle motion in a nonuniform electric field controlled by the insulating channel geometry, is a promising technique to study cell surface dielectric properties exclusively without introducing electrodes into the microchannel. To date, iDEP has been exploited to distinguish, characterize, and manipulate particles and bacteria based on their size and general cell phenotype (e.g. gram positive vs. gram negative). However, separation of bacteria with diverse surface polarizabilities but similar sizes sets a much higher demand on separation sensitivity, necessitating improvement in channel structure design in order to increase the electrical field gradient. In this work, a three dimensional insulator based dielectrophoresis (3DiDEP) microdevice is designed to achieve continuous particle sorting based on their size and, more precisely, on their surface dielectric properties. A 3D constriction is fabricated inside Poly (methyl methacrylate) (PMMA) channels using a micromilling technique. By controlling the channel geometry at the 3D constriction area, a nonuniform electric field with a large intensity gradient perpendicular to the local particle flow direction results in transverse particle deflection, driving particles into different outlet streams. With both simulation and experiments, we will show that a diverse array of particles and cells can be distinguished by the 3DiDEP device. This 3DiDEP sorter can be used in multiple applications in which the surface properties of cells and particles are of special interest.

(102c) Electrorotation As a Tool to Study Interaction Kinetics Between Proteins and Cells inReal-Time

Kilchenmann, S., EPFL

Spiga, F., EPFL

Guiducci, C., EPFL

Originally presented on: 11/4/2013 13:14:00 - 13:36:00

Electrorotation as a tool to study interaction kinetics between proteins and cells in real-time

Samuel Kilchenmann*, Fabio Spiga* and Carlotta Guiducci*

*École Polytechnique Fédérale de Lausanne, SWITZERLAND

Introduction

Electrorotation has been used to assess electrical parameters of biological particles such as cells, therefore allowing to distinguish between *e.g.* viable and non-viable yeast cells [1]. In biological sciences the evolution of cellular parameters over time is also very informative, for instance to determine the kinetics of protein-cell interactions in real-time. Furthermore, the possibility to simultaneously observe phenotypic changes of the cells allows to correlate them with the electrokinetic data. In the present work, we propose an electrorotation setup coupled with high resolution fluorescence imaging to observe the particle's movements. The bulky size and the short working distance of the conventional microscope lenses, which are normally used for high resolution fluorescence imaging, are critical for the fabrication of the electrorotation chips. For this reason, we developed a fabrication process that allows aligning microfluidic channels made in SU-8 with photolithographic accuracy with a glass top coverage, and featuring backside microfluidic accesses. Furthermore, these chips include 3D metal electrodes providing a high linearity of the rotating electric field, without any field component in the z-axis. These electrodes are advantageous to focus and follow single particles over time, since the particles are not moved out of the focal plane by dielectric forces resulting from the z-components of the electric field, as it commonly happens with planar electrodes.

Microfabrication

The fabrication of the electrorotation chips starts by defining the microfluidic accesses with standard photolithography and transferring the pattern 300 µm deep into a passivated silicon substrate. Then, the 3D structures are photopatterned in SU-8 that is afterwards covered with a metal layer by sputter deposition. Then, a dry film resist is applied and structured and it serves as etch mask for the metal layer. A further step of SU-8 lithography is done to define the microfluidic channels. In between the post exposure bake and the development of the SU-8, the resist topography is smoothened out by chemical mechanical polishing. After development of the second SU-8 layer, the wafers are grinded from the backside down to 300 µm in order to open up the microfluidic accesses. Finally, the wafers are diced and a microscopic cover slip is thermally bonded on top of the microfluidic channel.

Fluorescence imaging and electrorotation of healthy and apoptotic HEK cells

The suitability of our system for high resolution imaging was tested by imaging non-fixed healthy and apoptotic HEK cells. Apoptosis was induced through heat shock (45 minutes at 44°C) followed by a 2 hours recovery at 37°C. The cells were then washed and stained in a solution containing sucrose (8.5%), dextrose (0.3%) and Acridine Orange (50 µg/ml). The conductivity of this solution was adjusted to 0.1 mS by the addition of PBS. By imaging the cells, it was possible to distinguish between healthy cells, showing a roundish and clear nucleus (Figure 1a), and apoptotic cells, with deformed and non homogenously stained nucleus (Figure 1b). Electrorotation experiments were then performed by applying a signal with amplitude of 4-5 V_{p-p} at a frequency of 100 kHz. In these experiments it was possible to correlate the fluorescence imaging with the electrorotation data and to observe that healthy cells have a different sense of rotation with respect to apoptotic cells. In particular, apoptotic cells rotate in the direction of the electric field, while healthy cells rotate in the opposite direction. The observed rotation of the particles is fairly stable, making it possible to use this setup to study the evolution of the particle's parameters in real-time over a extended time.

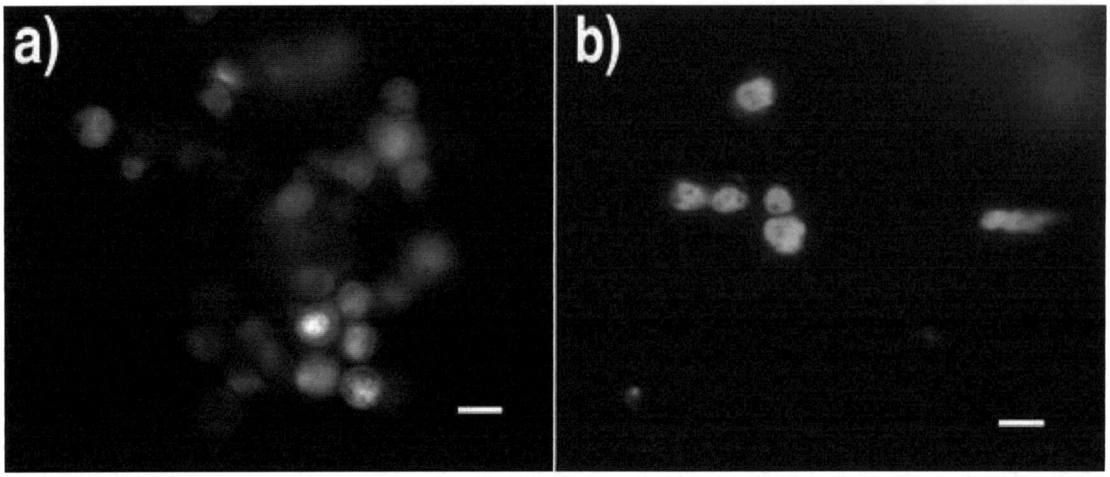

Figure 1: Fluorescence picture of normal (a) and apoptotic (b) HEK cells taken inside a microfluidic channel

1. Ying, H., et al., *Differences in the Ac Electrodynamics of Viable and Nonviable Yeast-Cells Determined through Combined Dielectrophoresis and Electrorotation Studies.* Physics in Medicine and Biology, 1992. 37(7): p. 1499-1517.

(102d) Studying the Effects of Sphingolipid Metabolites On Cells' Dielectrophoretic Properties Using Contactless Dielectrophoresis

Salmanzadeh, A., Virginia Tech
Savage, E., Virginia Tech
Roberts, P. C., Virginia Tech
Schmelz, E. M., Virginia Tech
Davalos, R., Virginia Polytechnic Institute and State University
Originally presented on: 11/4/2013 13:36:00 - 13:58:00

Current conventional cancer treatment regimens often rely upon toxic chemotherapeutics or target oncogenes that are variably expressed within the heterogeneous cell population of tumors. These challenges highlight the need for novel treatment strategies that can act on cells independently of variably expressed biomarkers and are non-toxic yet able to at least partially reverse the aggressive phenotype of the disease to a benign or very slow-growing state. To investigate the effect of non-toxic concentrations of sphingosine, a tumor-suppressor metabolite, and sphingosine-1-phosphate, a tumor-promoter metabolite, on the intrinsic electrical properties of mouse ovarian surface epithelial (MOSE) cancer cells, we used a label-independent rapid microfluidic cell manipulation platform known as contactless dielectrophoresis (cDEP). cDEP uses microfluidic electrode channels to provide a non-uniform electric field, avoiding direct contact between metal electrodes and the sample traditionally required by DEP. This contactless method eliminates electrochemical contamination, extensive sample preparation (no antibody labeling), and can maintain sterility of the sample. Previously, we used cDEP to demonstrate that electrical properties change as MOSE cells progress from a benign early stage (MOSE-E) to later malignant stages. Here, we demonstrate an association between So treatment and a shift in the bioelectrical traits of late stage MOSE (MOSE-L) cells towards a profile similar to that of benign MOSE-E cells. The specific membrane capacitance of MOSE-L cells shifted toward that of MOSE-E cells, decreasing from 23.94±2.75 to 16.46±0.62 mF/m^2 after So treatment. In contrast, S1P did not reverse the electrical properties of MOSE-L cells. This work demonstrates the potential for this microfluidic cDEP platform to be used as a new, rapid technique for drug efficacy studies, and eventually designing more personalized treatment regimens.

(102e) Dielectrophoretic Separation of Microorganisms Based On Speciation and Life-Cycle Stage

Su, Y. H., University of Virginia
Varhue, W., University of Virginia
Zaenker, E.,
Warren, C.,
Swami, N., University of Virginia
Originally presented on: 11/4/2013 13:58:00 - 14:20:00

Micro-organisms are usually present in heterogeneous samples that are spread over a range of developmental lifecycles and species type. Since antibiotic treatments are influenced by interactions across species type and developmental lifecycle stages, the development of antibiotic treatments can greatly benefit from techniques capable of separating microbial samples based on developmental lifecycle stage and species type. Herein, we investigate the application of dielectrophoresis towards accomplishing such separations for infectious organisms, such as *Cryptosporidium parvum*, a water-borne pathogenic protozoan parasite [1] and *Clostridium dificile*, a bacterial species implicated within hospital infections [2]. In parallel to their dielectrophoretic separation on a microfluidic device with insulator constrictions [3], the magnitude and direction of the trapping force is measured through real-time tracking of the velocity profile of single microbial cells for computing their dielectric properties based on fitting the frequency response of the force to a standard shell model. In this manner, dielectric properties of the microbial species can be correlated to their phenotype to optimize conditions, such as medium conductivity, and applied field and frequency towards enabling more effective separations. The application of this methodology is presented towards separating and characterizing the heterogeneous alterations of *Cryptosporidium parvum* after contact with disinfectants for quantifying the integrity of sporozoites and the oocyst wall to a high degree of sensitivity, in spite of the heterogeneous nature of their modification. Similarly this separation methodology is used to study the effect of interactions across species type (toxic versus non-toxic) and developmental lifecycle stages (spore versus vegetative) on the efficacy of antibiotic treatments on *Clostridium dificile.*

[1] Chen, X.M., et al., "Cryptosporidiosis". N Engl J Med, 2002. 346(22): p. 1723-31.

[2] M. Rupnik, M. H. Wilcox, D. N. Gerding. "*Clostridium dificile* infections: new developments in epidemiology and pathogenesis, *Nature Review*(2009), 7, 526-536.

[7] Swami, N.; Chou, C.-F.; Ramamurthy, V.; Chaurey, V. Enhancing DNA hybridization kinetics through constriction-based. *Lab on a Chip* **2009,** *9,* 3212–3220.

(102f) An Integrated Electrokinetic Chip Platform for Point-of-Care Nucleic Acid Diagnostics

Slouka, Z., University of Notre Dame
Senapati, S., University of Notre Dame
Chang, H. C., University of Notre Dame

Originally presented on: 11/4/2013 14:20:00 - 14:42:00

We present a reliable, sensitive, specific and disposable portable diagnostic biochip platform for DNA/RNA detection, that is based on a variety of modern electrokinetic phenomena/technology and can be powered by a low-voltage DC battery. This turn-key platform integrates a few indepenent units on a glass slide sized chip and is fully automated such that a sequence of multi-plex analysis is carried out without external interference. All units are economically fabricated with low-cost polymeric materials, including a microfluidic chip consisting of multiple layers of polycarbonate film, such that the overall material cost is less than one dollar per disposable chip. Despite its low cost, the multi-plex chip involves a sequence of intricate but robust operations that are connected by a continuous flow, with pumping providied by an on-chip pump, such that microfluidic flow can be used to process a large volume of real physiological sample, separate debris, eliminte diffusion transport limitation and provide shear-enhanced specificity. These units include: (i) a pretreatment unit with an electrophoretic gel loading component for extraction/separting RNA/DNA molecules from real samples with no or minimum upstream pretreatment(blood, saliva, urine etc), (ii) a depletion-zone preconcentration unit for localizing the DNA/RNA molecules next to the sensor, (iii) an ion exchange particle/membrane-based sensor, and (iv) an electroosmotic pump with a set of reservoirs for dosing the necessary buffers.

Chemically lysed samples are pipetted into the sample reservoir of the platform from which negatively charged biomolecules are loaded into the main channel through a thin agar gel by a DC electrical field. This loading process based on the principle of gel electrophoresis efficiently separates molecules of interests from cell debris and positively charged molecules. Loaded molecules are then electro-osmotically transported to the sensor/preconcentration unit which accumulates the negatively charged molecules at the ion exchange particle/membrane sensor, thus allowing specific target molecules to hybridize with probes on the sensor surface. The accumulation of negatively charged molecules in the high continuous flow (~10 microliter/min) at the sensor is achieved with a preconcentration membrane whose concentraton polarization effect introduces a depletion zone in the flow channel, which then acts as a filterless sieve for the target molecules.

The core of the diagnostic platform is a single, positively charged ion exchange particle/membrane used as a sensor. Ionic current through the ion exchange particle/membrane is sensitive to the presence of physisorbed DNA and RNA molecules due to surface charge inversion phenomenon caused by the negative charge of DNA/RNA phosphate-sugar backbone. This rather nonspecific detection mechanism of nucleic acids is rendered specific with covalent attachment of ssDNA oligoprobes complementary to the DNA/RNA targets. The ion current allows qunatification of the hybridized targets which are preferentiall retained with the non-targets being sheared away by the strong continuous flow. Current limit of detection is 1000 RNA molecules in a 10 microliter sample of raw sample, with a dynamic range from microM to fentoM or 10^3 to 10^9 targets and a specificity of two mismatched pairs for a 20b RNA.

We will present optimization/fabrication details for each unit of the detection platform and show results for the integrated platform for the detection of long single stranded mRNA and microRNA for E. Colli in water and oral cancer cells in saliva. We will also discuss future development of multi-target sensing chips.

(102g) A Novel Method for Measuring Proteins Zeta Potential Using Electroacoustics

Parlia, S.,
Dukhin, A., Dispersion Technology Inc
Originally presented on: 11/4/2013 14:42:00 - 15:04:00

It is known that measuring the zeta potential and valency of proteins in a physiological solution with high ionic strength presents a problem for traditional electrophoretic methods. High ionic strength leads to high current, thermal convection, and electrochemical contamination of the samples. This problem can be solved by using ultrasound waves instead of an electric field for moving protein molecules relative to the liquid. Propagation of ultrasound waves through a liquid containing charged entities creates an electric response called Colloid Vibration Current. Here we use a device that is capable of measuring this electroacoustic effect for characterizing the electric properties of several proteins dissolved in water. We present results for protein samples in different ionic compositions of water, as well as a high ionic strength physiological solution. We also investigate the role of temperature by conducting temperature sweeps from ambient temperature up to 36 °C.

(131a) Insulator Based Dielectrophoresis for the Manipulation of DNA Origami

Gan, L., Arizona State University
Chao, T. C., Arizona State University
Camacho, F., Arizona State University
Yan, H., Arizona State University
Ros, A., Arizona State University
Originally presented on: 11/4/2013 12:30:00 - 12:52:00

Self-assembled DNA nanostructures, such as DNA origamis have large potential for cutting-edge nanotechnological applications, such as DNA computing, photonic devices, or targeted diagnostics. The study of the properties for such nanostructures, controlled positioning as well as the concentration and purification is thus of utmost importance for their technological applications. In this work, we demonstrate manipulation and trapping of origami species with an insulator-based dielectrophoresis (iDEP) approach for the first time to the best of our knowledge. Specifically, we investigate the low frequency dielectrophoretic behavior of the 6-helix bundle origami in an iDEP device. We demonstrate a specific dependency of the trapping position on frequency together with a characteristic range of electric fields, in which origami DEP can be observed. We postulate that our findings can be directly applied to separation and purification of DNA origami species, as previously shown with linear DNA.

We designed an iDEP device in poly(dimethylsiloxane) to study 6-helix bundle DNA origami DEP. Via the application of potentials in the channel reservoirs, we create AC electric fields around the elliptic posts leading to electric field gradients large enough to trap the 6-helix bundle DNA origami. We observed that trapping occurs in a range of 200 V/cm to 2100 V/cm applied electric field spanning a frequency range of 60 Hz to 15 kHz. Interestingly, the frequency range under which trapping is observed is shifted to higher values for increasing applied electric field. Furthermore, the trapping area around the posts 'shrinks' with increasing frequency. To elucidate this phenomenon we performed numerical simulations with COMSOL Multiphysics. Our model solves the time dependent concentration distribution in the post array using the transport of diluted species module. We assume that the 6-helix bundle DNA origami is subject to diffusion, electrokinetic transport and DEP. The simulations result in a frequency dependent 'wiggling' migration which could be resolved experimentally due to the limitations in employed optics requiring large exposure times. However, we are able to compare the experimentally observed trapping area with the numerical simulations demonstrating that the experimentally determined distances match the theoretically predicted values in excellent agreement. This indicates that the physical origin of the variations in the trapping length are due to an overlay of electrokinetic and dielectrophoretic effects.

In summary, we demonstrated iDEP manipulation with the 6-helix bundle DNA origami. The characteristic frequency dependence on the dielectrophoretic trapping behavior was observed experimentally and substantiated with a numerical model. The variations in the trapping area around insulating structures can be explained by the electrokinetic migration of 6-helix bundle DNA origami within the trapping area. Our work provides a basis for the optimization of analytical techniques based on DEP such as fractionation and separation of origami species.

(131b) Bioparticle Differentiation in An Insulator-Based Dielectrophoretic Microchannel

Jones, P. V., Arizona State University
Hayes, M. A., Arizona State University
Originally presented on: 11/4/2013 12:52:00 - 13:14:00

A significant and wide-ranging body of work is focused on rapid bioparticle characterizations. Clinical diagnosis will be directly impacted by innovations in this field because it will enable rapid, on-site bioanalysis, improving the availability, accuracy, and scope for these tests. For bioparticle analysis, microfluidic electrokinetic approaches provide unique advantages, including short analysis times, microliter sample and reagent volumes, potentially low cost, and portability. The work presented here explores a separations-based strategy for isolating and concentrating intact microorganisms within a single, continuous microchannel.

Using DC insulator-based dielectrophoresis in a converging, sawtooth-patterned microchannel creates multiple and distinct bioparticles traps. This channel design enables localized isolation and concentration of a wide variety of microbes and other particles based characteristic differences in their physical properties. Various targets have been captured and concentrated within the device, including human blood cells and mature amyloid protein fibrils. Recent work has focused on using this approach to differentiate very similar targets, such as different strains of a single bacterial species. This approach has demonstrated the ability to distinguish three serotypes of live *Escherichia coli*, indicating the potential this technique holds in terms of both separatory power and diagnostic applications.

(131c) Fabric Microfluidics for Low-Cost Protein Separations

Narahari, T., Northeastern University
Dendukuri, D., Achira Labs Private Limited
Murthy, S., Northeastern University
Originally presented on: 11/4/2013 13:14:00 - 13:36:00

Introduction: Point-of-care diagnostic tests (POCTs) are necessary to screen for disease and monitor treatment at the site of patient care. By this definition, a POC device must be standalone, consume minimum laboratory resources and also be inexpensive [1]. For instance, the typical paper-based POCT for pregnancy consists of a designated path for a small volume of urine to flow along, complex with assay reagents and generate a result in terms of a colored signal that can be interpreted by the lay user. However, the manufacturing process involves multiple coating, drying and alignment steps that can limit scalability and contribute to cost. In prior work, we demonstrated a seamless, woven fabric-based microfluidics platform to address this issue [2]. Weaving allows us to combine reagent patterning and device assembly into a quasi-single step. Two main avenues will be discussed in the context of fabric POCTs: a. Passive control of liquid flow in fabric b. Feasibility of fabric as a platform for the electrophoretic separation of proteins from biological samples.

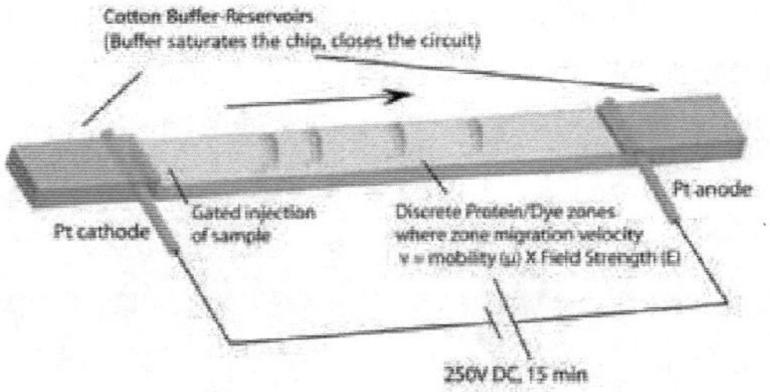

Figure 1: Schematic of the fabric electrophoresis device

Materials and Methods: Fabric samples were weaved on a handloom at Achira Labs, Bangalore, India. The flow properties that are required to maneuver medical samples such as serum or urine along a reagent-loaded pathway are provided by controlled structural in-homogeneities in the fabric. To manufacture the electrophoretic chip, Platinum wire electrodes were weaved into a polyester flow path with absorbent cotton ends as buffer-reservoirs (Figure 1).Â An Ohm's Law Plot was performed to ensure that the system could withstand a minimum of 250V DC. In order to discern diffusion effects from electrophoresis, dyes that can be tracked visually were run on fabric with and without an applied voltage. After confirming dye migration in an electric field, a dye-conjugated protein ladder (ColorBurst^{TM}, Sigma) was resolved in fabric at 250V and compared to a standard PAGE run.

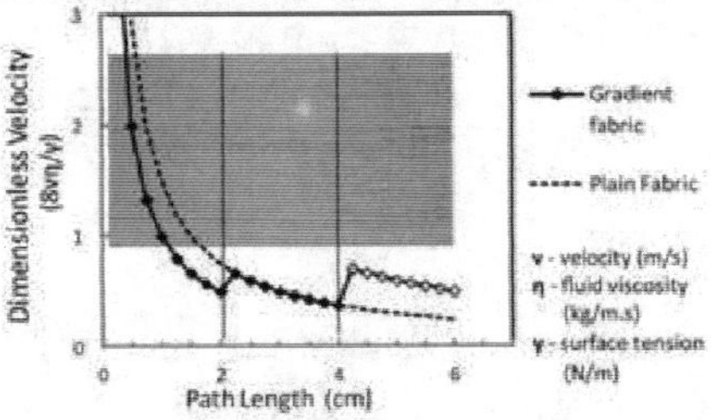

Figure 2: Velocity profile generated using aqueous buffer (1X PBS) in fabric with decreasing hydraulic resistance. Velocity profile in plain or homogeneous fabric is provided for comparison. (Dendukuri et. al., Achira Labs- unpublished work)

Results and Discussion: Figure 2 shows a saw-toothed velocity profile generated in a fabric where hydraulic resistance decreases in three seamless steps.Â This illustrates that sample flow can be slowed in regions where time-sensitive reactions must take place. Â The colored protein ladder was resolved by electrophoresis. Future work will attempt to resolve fluorescent proteins for use with the more accurate and quantitative detection of fluorescence.

Conclusions: From prior work [1] and from the electrophoretic separation achieved, fabric is a versatile platform for a variety of analytical procedures. Added to the high manufacturing throughput per-day on a standard industrial loom, fabric POCTs can be disseminated at low costs to resource deprived medical laboratories.

References:

1. P. Yager, G. J. Domingo, J. Gerdes (2008). *Annu. Rev. Biomed. Eng.* **10**: 107-144

2. P.Bhandari, T.Narahari, D.Dendukuri (2011). *Lab Chip* **11**(15): 2493-2499.

(131d) Microfluidic Approach for Antibiotic Susceptibility Testing of Polymicrobial Cultures

Mohan, R., University of Illinois, Urbana Champaign
Sanpitakseree, C., University of Illinois at Urbana Champaign
Sevgen, E., University of Illinois at Urbana-Champaign
Desai, A. V., University of Illinois at Urbana-Champaign
Schroeder, C. M., University of Illinois at Urbana-Champaign
Kenis, P. J. A., University of Illinois at Urbana-Champaign
Originally presented on: 11/4/2013 13:36:00 - 13:58:00

Presence of multiple bacteria in several clinically important infections (polymicrobial infections), such as urinary tract infections, peritonitis, abdominal sepsis, and aspiration pneumonia, are known to have higher mortality rate than monomicrobial infections such as tetanus or typhoid fever [1, 2]. Polymicrobial antibiotic susceptibility testing (AST) is not well studied, even though bacterial interactions are known to affect the AST outcome [3]. Hence, inappropriate antibiotic treatments in polymicrobial infections are often a major cause of mortality. Prior studies have relied on tedious methods such as quantitative terminal restriction fragment length polymorphism (qt-RFLP) and conventional plating methods that are time consuming and qualitative [4]. *Following our prior work on AST using E. coli as model pathogen [5], we now extend our microfluidic approach to quantify bacterial interactions in polymicrobial infections and determine susceptibility of bacteria in polymicrobial communities against commonly used antibiotics.*

Specifically, we utilized a 4x12-array multiplexed microfluidic platform for studying and quantifying antibiotic susceptibility of *Escherichia coli*, *Pseudomonas aeruginosa*, *Klebsiella pneumoniae*, *Acinetobacter baumanii* and *Burkholderia multivorans* in mixed cultures. Green and red fluorescent reporters were used to distinguish bacteria in a mixed culture and time lapse fluorescence microscopy was used for direct visualization and quantification of cell numbers over long periods of time. In a mixed culture of *P. aeruginosa* and *E. coli*, we observed lysis of *E. coli* and decreased growth rates of *P. aeruginosa* by approximately 20-40 minutes. In addition, we compare the minimum inhibitory concentration (MIC) of antibiotics against bacteria in mixed cultures with their constituent bacterial isolates. Interestingly, the MIC of antibiotics against *P. aeruginosa* increased drastically in presence of *E. coli* and *K. pneumoniae*. For example, the MIC of *P. aeruginosa* against tobramycin *increased* by a factor of 8 mixed with *E. coli*. In our ongoing work, we are performing similar experiments with bacteria including *A. baumannii* and *B. multivorans*.

This microfluidic approach offers several advantages such as rapid turn-around times (2-4 hours vs. 2-4 days), reduced sample volumes (~ 2 nL vs. 30 mL), and high detection sensitivity (1 cell *vs.* 10^6 cells) over current assay methods. Most importantly, our results prove that significant differences exist in susceptibility of bacteria in polymicrobial cultures *vs.* monomicrobial cultures that can be rapidly diagnosed and studied using our unique diagnostic platform for a largely understudied but clinically important field of polymicrobial infections.

[1] F. Heilmann, *Infection*, **1993**, 3, 187-190.
[2] A.R. Marra *et al.*, *BMC Infectious Diseases*, **2005**, 5, 1471-1565
[3] C. Riedele and U. Reichl, *Eng Life Sci.*, **2012**, 12,188-97
[4] J.K. Schmidt *et al.*, *Biotechnol Bioeng*, **2007**, 96, 738-56
[5] R Mohan *et al.*, *Biosensors and Bioelectronics*, **2013** (Accepted)

(131e) A Low-Cost Nucleic Acid Biosensor for Point-of-Care Application

Senapati, S., University of Notre Dame
Slouka, Z., University of Notre Dame
Shah, S., University of Notre Dame
Chang, H. C., University of Notre Dame
Originally presented on: 11/4/2013 13:58:00 - 14:20:00

A Low-Cost Nucleic Acid Biosensor for Point-of-Care Application

Satyajyoti Senapati, Zdenek Slouka, Sunny Shah, and Hsueh-Chia Chang

Department of Chemical and Biomolecular Engineering

University of Notre Dame, Indiana 46556

Abstract

The development of low-cost and portable detection platform for genetic identification has attracted great attention to a wide variety of application such as infectious disease control, epidemic control, clinical diagnosis etc. Nucleic acid amplification-based detection systems (real time PCR) specifically designed for portability are far from reality due to their complex protocol, difficulty of use and high costs limiting their widespread use among first responders and public health officials. Alternatively, a simple label-free and PCR-free turn-key sensing platform based upon direct nucleic acid (DNA/RNA) detection would have the potential to identify the specific virus and strain as well as detect infection in patients more quickly and specifically than immunoassays. In our prior studies, we discovered a nanoporous ion-selective sensing platform that can achieve these objectives. Here, we report a low-cost, portable, selective and PCR free biosensing platform for the detection of DNA/RNA molecule using ion selective environments. The detection is based on a charge inversion phenomenon that occurs when negatively charged DNA/RNA molecules hybridize with specific probes attached to ion exchange nanoporous materials bearing a positive fixed charge. When a DC field is applied across a nanoporous material, a small molecular charge polarized layer develops on the depletion side after concentration polarization. The presence of molecules in this polarized layer can hence sensitively change the limiting and overlimiting currents of the ion-selective material. Furthermore, oppositely charged small molecule targets can produce a bipolar membrane, like a P-N semiconductor junction, that can produce a large water-splitting signal when a high field develops at the carrier-depleted junction with a reverse bias. This new molecular sensing technology also accelerates the assay time because the high field in the polarized layer can attract large nucleic acids by dielectrophoresis and the depletion region can concentrate the charged molecules near the sensor surface in a flowing stream. Instead of the classical electrochemical sensing technique, we utilize a much more sensitive conductance and impedance signature based on sensitivity of ion current across the nanoporous membrane or particle to surface charge density change effected by hybridized molecules. The sensitivity involves a nonlinear correction to the conductance and is hence much more sensitive than EIS or other linear electrochemical sensing techniques.

(131f) Pico-Force Optical Exchange (pico-FOX): Separation of Particles From Molecular Components Utilizing Optical Forces With Orthogonal Fluid Flow With Applications to Malaria

Staton, S. J. R., Arizona State University
Kim, S. Y., National Research Council
Hebert, C.,
Hart, S., Naval Research Laboratory
Collins, G., Naval Research Laboratory
Terray, A., Naval Research Laboratory
Originally presented on: 11/4/2013 14:20:00 - 14:42:00

Presented here are the results of a combined pico-Newton optical force applied orthogonally to a hydrodynamically pinched mixed sample of molecular and particulate components. The pico-FOX microfluidic platform successfully combines optical forces and electrokinetic forces to promote higher resolution separations by capitalizing on the different properties of a complex sample mixture. The combined portfolio of pooled separation vectors of the integrated system increases the diversity of exploitable particle properties, including: size, shape, refractive index, charge, charge distribution, charge mobility, permittivity, and deformability. The pico-FOX microfluidic platform consists of a simple 2-D fused silica constructed microfluidic device with a symmetric inlet/outlet system that hydrodynamically pinches the mixed sample flow. The simplicity of this geometry enables the separation of particulates into resolved sub-groups that are simultaneously removed from contaminating molecular components. Particle separations were achieved by addressing particles to specific fluid lamina that flowed into different outlet effluents. The pico-FOX device has been applied to a variety of both abiotic and biotic particulate types.

Experiments utilizing analog abiotic particles have demonstrated non-invasive separations based on single property differences such as size, refractive index, and electrophoretic mobility. Particle separations tested the different separation modes available, from purely optical force to combined optical and electrophoretic forces. These experiments utilized different combinations of polystyrene, PMMA, and silica particles with a commercially available dye. The particle properties varied in size (2, 6, 10 μm), refractive index, and electrophoretic mobility. Sized based separations of 6 and 10 μm polystyrene particles from a mixed component sample led to only 3% particle contamination in the original sample stream and inter-particle type enrichment levels >80% in their respective outlet effluents.

Separations exploiting differences in refractive index and electrophoretic mobility were able to uniquely address three different particle materials (polystyrene, PMMA, and silica) of the same size (2 μm) to a unique outlet without measureable contamination of the other particle types. Particles with the largest refractive indices experience the largest amount of displacement due their interaction with the laser. This was observed in the separation of PMMA, =1.49, from silica, =1.43, where PMMA was separated to a more distal outlet than the silica. [1] Although polystyrene has the highest refractive index, =1.59, and would normally be expected to experience the largest optical force, under the experimental conditions utilized, it never interacted with the laser due to its electrophoretic mobility pulling it in the opposite direction. Given that all of the particles are the same size and experience the same experimental conditions, the variation in electrophoretic mobility of the different particles in the system can be reduced to differences in zeta potential (ζ), or the electrical potential of the interfacial surface double layer. Polystyrene has the largest absolute ζ of the three different particle types, causing the polystyrene particles to be the first to respond in an electrophoretic manner. The reported ζ of 2 μm polystyrene particles is -70.8 mV, 2 μm silica particles is -60 mV, and 2 μm PMMA particles is +34.9 mV.[2-4] Thus, the pico-FOX separation method combines electrophoretic and optical forces to distinguish the three different particle types.

In the testing of analog abiotic particles, pico-FOX has demonstrated a separating size resolution of approximately 4 μm. Also promising is that although the particle flow densities were not optimized for maximum particle processing in these tests, the average processing rate reached over 5,500 particles per hour. In addition to particle separation, the device was able to minimize dye diffusion, leading to > 95% dye recovery. For all sample types examined, the resolution of the separation is dependent on the width of the hydrodynamically pinched sample flow along the wall at the point that the sample flow interacts with the laser beam. Given that the particles travel the same distance when displaced by the optical force, the width of the pinched flow is preserved in the final separated sample. Balancing dilution rate from increased pinch flow and separation resolution, the width of the pinched sample flow was ~50 μm. Modeling and optimization of the pico-FOX system, its separation potential, and theoretical underpinnings have been described and formed the basis for this experimental investigation. Theoretical calculations were conducted using COMSOL, a finite element multiphysics model, for geometric optimization and prediction of separation performance. Experimental separation performance for particulates and molecular component are in good agreement with modeled performance.[5]

Applications of pico-FOX on the separation of biological particulates were evaluated using malaria infected red blood cells and other blood components. Separation of blood components were investigated primarily through size variations in the main classes of blood components. However, the separation of red blood cells infected with

Plasmodium falciparum, one species of protozoan that causes malaria, explores separating infected versus non-infected cells via differences in birefringence. Malaria infected red blood cells respond to linearly polarized laser light via intrinsic birefringence by rotating axially along the propagation of the beam. This is in contrast to non-infected red blood cells which are only form birefringent, causing them to simply align along the beam. Pursuing separations utilizing variations in birefringence in an orthogonal flow through system would enable the high throughput necessary to process sufficient quantities of red blood cells to address the low levels of parasitemia found in the different types of malaria that can range from 2 to >50% of red blood cells.[6]

Potential applications of this technology in the fields of biological and environmental sciences are promising. Pico-FOX is capable of removing particulates from potentially damaging molecular components in a mixed sample. This ability could prove useful in biomedical applications that require sample pretreatment prior to analysis steps, such as total sample analysis on a microfluidic platform. In settings where duplicative sampling would be prohibitive, such as medical cell sorting, bio-threat agent detection, water management and pollution detection, this technology could be used to treat a single, complex sample in a rapid and efficient manner. This sample could then be mated to a variety of different analytical detection techniques to analyze the various constituents. The capability for efficiently separating molecular components from particulates is especially valuable for samples containing a mixture of bacteria, protozoan, etc. and molecular components, e.g. chemical warfare agents, pollutants, and toxins. Two such common water-borne pathogenic protozoa are *Cryptosporidium parvum* and *Giardia intestinalis,* nominally 5 and 10 μm, respectively, which are comparable to the sizes tested here.[7] A non-invasive and non-destructive testing device tailored to particular analytes could find broad applications in water management and malaria detection, considering that annual deaths worldwide for all types of waterborne pathogens total more than 3.4 million and for malaria number around 660,000.[6, 8] In future applications, modeling of the desired particulates would aid in tailoring the geometry of the chip to offer the appropriate outlet configurations to maximize enrichment of the different sub-samples via their optical and electrophoretic properties. The combination of the non-invasive optical forces and electrokinetics is a rich area for improving complex sample separation and manipulation.

References:

1. Hart, S. J.; Terray, A. V., Refractive-index-driven separation of colloidal polymer particles using optical chromatography. *Applied Physics Letters* **2003,** *83*, (25), 5316-5318.

2. Tabata, Y.; Ikada, Y., Effect of the size and surface charge of polymer microspheres on their phagocytisis by macrophoage. *Biomaterials* **1988,** *9*, 356-362.

3. Kim, J.; Lawler, D. F., Characteristics of Zeta Potential Distribution in Silica Particles. *Bulletin to the Korean Chemical Society* **2005,** *26*, (7), 1083-1089.

4. Pham, V. H.; Dang, T. T.; Hur, S. H.; Kim, E. J.; Chung, J. S., Highly Conductive Poly(methyl methacrylate) (PMMA)-Reduced Graphene Oxide Composite Prepared by Self-Assembly of PMMA Latex and Graphene Oxide through Electrostatic Interaction. *Applied Materials and Interfaces* **2012,** *4*, 2630-2636.

5. Staton, S. J. R.; Terray, A.; Collins, G. E.; Hart, S. J., Orthogonal optical force separation simulation of particle and molecular species mixtures under direct current electroosmotic driven flow for applications in biological sample preparation. *Electrophoresis* **2013,** *34*, 1175-1181.

6. Trampuz, A.; Jereb, M.; Muzlovic, I.; Prabhu, R. M., Clinical review: Severe malaria. *Critical Care* **2003,** *7*, 315-323.

7. *Foodborne Parasites*. Springer: New York, 2006.

8. Berman, J. WHO: Waterborne Disease is World's Leading Killer. (Feb. 6), 2009.

(131g) Single Cell Analysis Using Time-Resolved Spectroscopic Measurements in Flow Cytometry

Houston, J. P., New Mexico State University
Li, W., New Mexico State University
Originally presented on: 11/4/2013 14:42:00 - 15:04:00

Flow cytometry is a fluidic-based bioanalytical tool in which single cells are observed for the total amounts of fluorescence and Raleigh scatter light that are emitted subsequent to laser excitation. As cells flow past the laser excitation source, the spectral-based measurements are rapidly made and interpreted to indicate new cellular phenomena and responses to treatments. Transient cell movement in flow cytometry allows high-throughput measurements but also typically limits the ability to carefully discriminate the distinct spectral details elicited from any given endogenous or exogenous fluorophore. In this contribution, we present a new time-resolved cytometry approach based on a time-domain spectroscopic approach. A rapidly scanning laser beam is controlled by an acousto-optic deflector, to interact with a cell multiple times as it rapidly moves at a laminar flow rate through a flow channel. Using deconvolution and other analysis approaches, the fluorescence decay is extracted. Fluorophores with different lifetime values ranging approximately from 3 ns to 25 ns have been examined with this new type of flow cytometer. The lifetime decay profiles appropriately reflect if the fluorophore has a longer fluorescence lifetime (~20ns) by having a longer "tail" or a shorter lifetime (~3 ns) by a shorter "tail." The instrument is further characterized by introducing multiple fluorescence lifetimes so as to observe multiple-fluorescence decays. This new concept of flow cytometry can be largely impacting in that it could lead to in-depth single cell analysis for rare events (i.e. circulating tumor cells), and alleviate widespread cytometry issues with spectral overlap.

(134a) Identifying Defect Sites in Lithium-Ion Battery Materials: Local Disorder in LiVPO4F and Its Influence On Bulk Properties

Messinger, R. J., CNRS
MÃ©nÃ©trier, M., CNRS
Carlier, D.,
Ateba Mba, J. M., CNRS
Croguennec, L.,
Masquelier, C.,
Massiot, D.,
Deschamps, M.,
Originally presented on: 11/4/2013 12:30:00 - 12:45:00

The transition metal fluorophosphate $LiVPO_4F$ is a promising electrode material for lithium-ion batteries that is capable of operating both as a high-voltage positive electrode (4.25 V, $V^{III/IV}$) and a negative electrode (1.8 V, $V^{II/III}$).[1,2] While its Tavorite-like structure has been solved by a combination of X-ray and neutron diffraction, solid-state nuclear magnetic resonance (NMR) measurements reveal a significant extent of local disorder that has not been observed at longer length scales with scattering techniques. In particular, solid-state single-pulse ^7Li NMR spectra acquired under conditions of ultra-fast (64 kHz) magic-angle-spinning (MAS) reveal that $LiVPO_4F$ exhibits several additional ^7Li NMR signals of unknown origin that can account for up to 20% of the total lithium content. Electrochemical measurements in Swagelok-type cells with lithium metal as the negative electrode establish that materials with different populations of unknown lithium sites exhibit different cycling behavior, including differing extents of irreversible capacity and polarization. 2D $^7Li\{^7Li\}$ dipolar correlation NMR measurements establish the sub-nanometer proximities of the unknown lithium sites with respect to the $LiVPO_4F$ site, proving unambiguously that the unknown sites are local defects within the $LiVPO_4F$ structure and are not associated with separate impurity phases. In combination with magnetic measurements, variable-temperature NMR measurements reveal that the paramagnetic ^7Li NMR shifts of the different lithium sites vary linearly with the bulk magnetic susceptibility, further confirming that the sites are defects within the $LiVPO_4F$ structure. The results provide a general strategy for identifying lithium defect sites in lithium-ion battery materials and highlight the importance of understanding local disorder and its relationship to macroscopic material and device properties.

[1]J.-M. Ateba Mba, C. Masquelier, E. Suard, L. Croguennec, *Chem. Mater.* **2012**, *24*, 1223 – 1234.

[2]J.-M. Ateba Mba, L. Croguennec, N.I. Basir, J. Barker, C. Masquelier, *J. Electrochem. Soc.* **2012**, *159*, A1171-A1175.

(134b) Electrochemical Properties and Characterization of Li1.2Mn0.54Co0.13Ni0.13O2 Composite Cathode Powders Prepared by Ultrasonic Spray Pyrolysis

Langrock, A., University of Maryland
Ehrman, S. H., University of Maryland
Wang, C., University of Maryland
Originally presented on: 11/4/2013 12:45:00 - 13:00:00

Lithium-rich layered oxide $Li_{1.2}Mn_{0.54}Co_{0.13}Ni_{0.13}O_2$ (LMO) composite cathode powders are prepared by ultrasonic spray pyrolysis. The crystal structure and morphology of the prepared samples are characterized with X-ray diffraction and scanning electron microscope. The spherical shape and porous nanostructure of the as prepared powders are maintained after post-treatment at temperatures between 600 and 800°C. Grain growth in the powders increases with increasing post-treatment temperatures. The BET surface area of the composite cathode powders post-treated at 800°C is 12 m^2 g^{-1}. Discharge capacity of the cathode powders post-treated at 800°C decreases from 278 mAh g^{-1} to 208 mAh g^{-1} by the 50th cycle, the capacity retention being 75%. The post-treatment temperature strongly affected the electrochemical properties of the composite cathode powders. Composite cathode powders post-treated at high temperatures with high amounts of the Li_2MnO_3 phase had good cycle properties as well as large initial charge/discharge capacities.

(134c) Atomic Layer Deposition (ALD) On the Nanostructured Li-Mn-Rich Composite Li1.2Ni0.13Mn0.54Co0.13O2 Cathode Powder

Zhang, X., Institute of Nuclear and New Energy Technology, Tsinghua University
Belharouak, I.,
Originally presented on: 11/4/2013 13:00:00 - 13:15:00

Atomic Layer Deposition (ALD) on the Nanostructured Li-Mn-rich Composite $Li_{1.2}Ni_{0.13}Mn_{0.54}Co_{0.13}O_2$Cathode Powder

Xiaofeng Zhang,[a] Ilias Belharouak,[a,*]

[a] Chemical Sciences and Engineering Division, Argonne National Laboratory, 9700 S. Cass Ave, Argonne, IL 60439, USA, Email: belharouak@anl.gov(Corresponding author)

Li-Mn-rich (LMR) composite cathode materials $Li_{1+x}M_{1-x}O_2$ (M = Ni, Mn and Co) are of particular interest for plug-in hybrid electric vehicles (PHEVs), in which a high-energy battery is always desired for a higher mileage per charge.[1, 2] In the composite structure, the layered, the spinel and rocksalt structure are compatible at nano-scales, that many composites with different electrochemical properties can be synthesized by simply tuning the Li content in the materials. However, these materials would suffer from voltage decay and capacity fading with cycles, which significantly deteriorate their energy densities and life time for practical applications. In this type of materials, Li content is essential to the arrangement of the transition metal atoms in the composites and also to the structural stability against undesired reactions in a real electrochemical cell. It has been demonstrated that the occurrence of layered-spinel transformation induces the voltage decay. It is because the Li intercalation/deintercalation reactions may preferentially occurs at the tetrahedral site other than octahedral site, which facilitates the spinel formation.[1, 2] Further study by HR-TEM indicated that the composite particle surface could lose the layered feature after a few cycles, and the first-principle calculation suggested a defect-spinel structure forming on the surface due to the migration of transition metal ions into the Li layers.[3] Therefore, the surface chemistry and structure stability appear to be essential to limit the undesired reactions and phase transformation in the composite materials. Some achievements have been made *via* developing new synthesis processes, new chemistry or surface modifications (e.g., coating), which possibly altered the short-range and long-range ordering of transition metal atoms (Ni and Mn) near the surface, which lead to an improvement in the battery performance.[4-8]

Atomic layer deposition (ALD) is a well-established method to grow ultrathin films on flat surfaces and powders in atomic thickness using sequential, self-limiting surface reactions.[9-14] It has been demonstrated that the surface film can react as a surface barrier between the cathode active material and the electrolyte, which would reduce the loss of active materials in the electrode. [9-14] Consequently, less capacity loss is achieved for the coated cells as compared with the pristine materials. The well-established ALD process normally is using trimethylaluminium (TMA) and H_2O as precursors forming an Al_2O_3 surface film. Experimental results showed that the coated $LiCoO_2$ powders exhibited a capacity retention of 89% after 120 charge–discharge cycles in the 3.3–4.5 V vs. Li/Li+ range. [12, 15-17] The ALD surface coating can also be applied to battery electrodes at a low temperature <150 °C, which would not degrade the electrodes (e.g., melting of polymer binders) in this temperature range. It is reported that Al_2O_3-coated $LiCoO_2$ electrodes display superior electrochemical performance in comparison to the uncoated $LiCoO_2$ electrodes. However, Liu et al observed that the ALD surface coated Al_2O_3 could be lithiated forming passivation $LiAlO_2$ glass on the materials surface using in-situ TEM [18], which provided new insight into the electrochemical characteristic of the Al_2O_3surface film against Li.

In this paper, we are aiming for studying the importance of a conformal surface protection film using atomic layer deposition (ALD) technique on the high-capacity $Li_{1.2}Ni_{0.13}Mn_{0.54}Co_{0.13}O_2$ cathode materials in Li-ion batteries, which can be potentially significantly improved in their electrochemical performance. More precisely, we evaluated the effect of coating species, coating thickness and morphology on the rate performance, cycling life and high temperature performance of the high-capacity $Li_{1.2}Ni_{0.13}Mn_{0.54}Co_{0.13}O_2$ cathode materials in half-cell and full-cell configurations.[19] We found that an conformal Al_2O_3 thin film could be deposited on the powders using 10 ALD cycles of TMA and H_2O at 150 °C. TiO_2 coating was prepared using 20 ALD cycles of titanium isopropoxide (TTIP) and H_2O at 150 °C, which displayed a particulate morphology. Nevertheless, the surface coated TiO_2 film was electrochemically active, which could be lithiated forming Li_xTiO_2 as detected by X-ray photoelectron spectroscopy (XPS). The electrochemical reactivity of TiO_2 will cause degradation of the surface film, leading to capacity fading of the cell. Nevertheless, it was also found that the Al_2O_3 surface coating was not electrochemically active after repetitive cycles, which was contrary to the early report by Lui et al that Al_2O_3 can be lithiated into $LiAlO_2$. [18] To further explorate the phenomena, a systematic study on the ALD coating of Al_2O_3 and $LiAlO_2$was performed.

Refereances:

[1] C.S. Johnson, N. Li, J.T. Vaughey, S.A. Hackney, M.M. Thackeray, Electrochem. Commun., 7 (2005) 528-536.

[2] C.S. Johnson, N.C. Li, C. Lefief, J.T. Vaughey, M.M. Thackeray, Chem. Mater., 20 (2008) 6095-6106.

[3] B. Xu, C.R. Fell, M.F. Chi, Y.S. Meng, Energy & Environmental Science, 4 (2011) 2223-2233.

[4] J. Liu, B. Reeja-Jayan, A. Manthiram, Journal of Physical Chemistry C, 114 (2010) 9528-9533.

[5] S.H. Kang, M.M. Thackeray, J. Electrochem. Soc., 155 (2008) A269-A275.

[6] J.R. Croy, M. Balasubramaniam, D. Kim, S.-H. Kang, M.M. Thackeray, Chem. Mater., 23 (2012) 5415-5424.

[7] J.R. Croy, S.H. Kang, M. Balasubramanian, M.M. Thackeray, Electrochem. Commun., 13 (2011) 1063-1066.

[8] D. Kim, S.H. Kang, M. Balasubramanian, C.S. Johnson, Electrochem. Commun., 12 (2010) 1618-1621.

[9] M.Q. Snyder, S. Trebukhova, B. Ravdel, M.C. Wheeler, J. DiCarlo, C.P. Tripp, W.J. DeSisto, Mater Res Soc Symp P, 972 (2006) 331-338.

[10] M.Q. Snyder, S.A. Trebukhova, B. Ravdel, M.C. Wheeler, J. DiCarlo, C.P. Tripp, W.J. DeSisto, J. Power Sources, 165 (2007) 379-385.

[11] S.M. George, Chem. Rev., 110 (2010) 111-131.

[12] Y.S. Jung, A.S. Cavanagh, A.C. Dillon, M.D. Groner, S.M. George, S.H. Lee, J. Electrochem. Soc., 157 (2010) A75-A81.

[13] L.A. Riley, A.S. Cavanagh, S.M. George, S.H. Lee, A.C. Dillon, Electrochem. Solid-State Lett., 14 (2011) A29-A31.

[14] L.A. Riley, S. Van Ana, A.S. Cavanagh, Y.F. Yan, S.M. George, P. Liu, A.C. Dillon, S.H. Lee, J. Power Sources, 196 (2011) 3317-3324.

[15] Y.S. Jung, A.S. Cavanagh, L.A. Riley, S.H. Kang, A.C. Dillon, M.D. Groner, S.M. George, S.H. Lee, Advanced Materials, 22 (2010) 2172-2176.

[16] I.D. Scott, Y.S. Jung, A.S. Cavanagh, Y.F. An, A.C. Dillon, S.M. George, S.H. Lee, Nano Lett., 11 (2011) 414-418.

[17] Y.S. Jung, P. Lu, A.S. Cavanagh, C. Ban, G.H. Kim, S.H. Lee, S.M. George, S.J. Harris, A.C. Dillon, Advanced Energy Materials, 3 (2012) 213-219.

[18] Y. Liu, N.S. Hudak, D.L. Huber, S.J. Limmer, J.P. Sullivan, J.Y. Huang, Nano Lett., 11 (2011) 4188–4194.

[19] X. Zhang, I. Belharouak, L. Li, Y. Lei, J.W. Elam, A. Nie, X. Chen, R. Shahbazian-Yassar, R.L. Axelbaum, Advanced Energy Materials, (2013). 10.1002/aenm.201300269

(134d) Metal Fluoride/Graphene Composites for High-Performance Li-Ion Cathodes

Hayner, C. M., Northwestern University
Zhao, X., Northwestern University
Yu, Y. Y., Northwestern University
Kung, M. C., Northwestern University
Kung, H. H., Northwestern University
Originally presented on: 11/4/2013 13:15:00 - 13:30:00

Rechargeable batteries with improved energy and power density have many potential, high impact applications including advanced portable electronics and electric vehicles. However, most 'next-generation' materials capable of providing improved performance suffer from rapid capacity degradation or severe loss of capacity when rapidly discharged. Graphene, a unique two-dimensional material composed of atomically thin carbon sheets, has shown potential to address unsatisfactory rate capability, limited cycling performance and abrupt failure of these high capacity material. Therefore, we have been exploring graphene-based architectures, allowing us to exploit graphene's unique combination of high surface area, high in-plane electrical conductivity, excellent tensile modulus and mechanical durability.

Metal fluorides, including iron trifluoride (FeF_3), have the potential of both high energy delivery and high cyclability. We have recently developed a stable nanostructured iron fluoride (FeF_3)/graphene composite cathode. A facile route that combines co-assembly and photothermal reduction was developed to synthesize free-standing, flexible FeF_3/graphene papers. The papers contained a uniform dispersion of FeF_3 nanoparticles (< 40 nm) and open ion diffusion channels in the porous, conducting network of graphene sheets that resulted in a flexible paper cathode with high charge storage capacity, rate, and cycling performance, without the need for other carbon additives or binder. The composites demonstrate promising applications as cathodes in high-energy density Li-ion batteries. In a free-standing form, they showed a high storage capacity of >400 mAh/g and improved cycling performance compared to bare FeF_3 particles. In addition, we have investigated a novel mixed metal fluoride cathode system, containing tunable manganese and iron fluoride, for which characterization and electrochemical performance will be presented.

(134e) Nanostructured Li2S-Graphene Composites As Cathode for High Energy Density Li-S Batteries

Han, K., Northwestern University
Shen, J., Northwestern University
Hayner, C. M., Northwestern University
Yu, Y. Y., Northwestern University
Kung, M. C., Northwestern University
Kung, H. H., Northwestern University
Originally presented on: 11/4/2013 13:30:00 - 13:45:00

Lithium-ion batteries, although a promising technology for energy storage, is reaching their energy density limitation, making it difficult to fulfill the requirements of emerging applications such as electric vehicles and smart grid. Lithium-sulfur batteries are one of the most attractive systems for next generation high energy storage beyond Li-ion batteries. Conventional Li-S batteries mainly use elemental sulfur as the starting cathode material due to its high theoretical capacity of ~1675 mAh/g and high energy density of ~2500 Wh/kg based on the redox reaction $S_8 + 16Li \rightleftharpoons 8Li_2S$. Sulfur is also low cost, abundant, and of low toxicity. However, significant hurdles such as rapid capacity fade and poor rate capability have to be overcome before practical application can be realized. Two important underlying reasons for the poor performance of this battery are the insulating nature of sulfur and the dissolution of the intermediate product polysulfide into the electrolyte resulting in redox shuttling. Although extensive research efforts have been expanded to solve these issues, most studies are limited to the simple cell configuration of sulfur as cathode and metallic lithium as anode. For safety and ease of operation, introducing Li into the cell in the form of Li_2S is more desirable than Li metal. Moreover, Li_2S, with a theoretical capacity of ~1166 mAh/g, could be paired with a lithium-metal-free anode such as Si or Sn.

Here, we report the use of a nanostructured Li_2S-graphene composite as cathode material. Li_2S nanoparticles (20~50nm) are imbedded in pocket-like structures in a three-dimensional graphene/reduced graphene oxide via a facile, wet chemical method followed by thermal treatment. Structural characterization by XRD and SEM provide clear evidence for the unique structure. The constructed 3-D graphene network could enable efficient use of Li_2S active material by providing excellent electrical contacts, offer long-term stability against loss of polysulfide and accommodate volume changes during charge and discharge processes. The as-prepared Li_2S-graphene nanocomposites exhibit high initial reversible capacity and good cycling performance. The effect of lithium polysulfide as additive of the electrolyte is also investigated, especially with respect to its contribution to the overall charge storage capacity. The structural and electrochemical characterization results of the nanocomposites will be presented.

Nanomaterials for Energy Applications

(134f) Carbon/Sulfur Microspheres With Multi-Modal Pore Structures for Lithium-Sulfur Battery Cathodes

Zhao, C., University of Wisconsin-Milwaukee

Liu, L., University of Wisconsin-Milwaukee

Zhao, H., University of Wisconsin-Milwaukee

Li, Y., University of Wisconsin-Milwaukee

Originally presented on: 11/4/2013 13:45:00 - 14:00:00

Sulfur has received increasing attention as a cathode material for lithium-sulfur batteries due to its high theoretical specific capacity of 1675 mAh g-1, which is five times higher than that of conventional lithium-ion batteries. However, poor electronic conductivity of sulfur, dissolution of the lithium polysulfide intermediates and large volume expansion of sulfur during cycling limit the commercialization of Li-S batteries. In this work, we have demonstrated a cathode design with a novel structure made of multi-modal porous carbon microspheres. The tunable porous structure of sulfur-infiltrated carbon nanocomposites allows easy access to Li ion transport, inhibits loss of active sulfur material, and provides void space for sulfur expansion. An initial capacity of 1289 mAh g-1 at 1C and a Coulombic efficiency of 90% over 200 cycles are achieved.

(134g) Nanofiber-Based Cathode Electrocatalysts Supported On Carbon Substrates for Lithium-Air Rechargeable Batteries

Joo, Y. L., Cornell University
Yin, J., Cornell University
Kim, J.,
Originally presented on: 11/4/2013 14:00:00 - 14:15:00

Nanofiber-Based Cathode Electrocatalysts Supported on Carbon Substrates for Lithium-Air Rechargeable Batteries

*Jun Yin[a], Jangwoo Kim[a], and Yong Lak Joo[a]**

[a] School of Chemical and Biomolecular Engineering, Cornell University, Ithaca, New York 14853

In terms of usable energy density rivaling that of gasoline, rechargeable lithium-air batteries have attracted increasing interests presently. However, the practical capacity was highly affected by the effective catalyst materials in the air cathode in Li-air batteries. The demand for effective catalyst materials in the air cathode requires the design and synthesis multifunction catalysts for power density improvement. In this presentation, several types of metal oxide nanofibers (e.g., ZnO, Co_3O_4 etc) were synthesized by water-based electrospinning of highly loaded precursor solutions, followed by thermal treatment and their use as the cathode catalysts in single-cell measurements was described. In the meantime, the correlation between the various carbon-based substrates and the metal oxide fiber catalysts and how they synergistically influence the electrocatalytic activity will also be discussed. The metal oxide nanofibers are characterized using an array of techniques such as SEM, XRD, XPS, BET and Raman to determine their morphology and structural properties. Charging – discharging evaluation, cyclic voltammetric measurements in organic electrolytes were carried out to assess the mechanistic details as well.

(134h) Low-Temperature Dehydrogenation From Aminoborane Complexes Under Carbon Dioxide Atmospheres

Lee, J. W., The City College of New York
Xiong, R., CCNY
Zhang, J., The City College of New York
Originally presented on: 11/4/2013 14:15:00 - 14:30:00

Ammonia borane (AB) is a chemical hydride and one of attractive solid-state media for on-board hydrogen storage due to high hydrogen contents (19.6 wt. %) and low-temperature dehydrogenation compared to other metal hydrides. The hydrogen release from AB can be accelerated at the working temperature (~ 85 °C) of polymer electrolyte membrane fuel cell. However, this talk shows that CO_2-treatment can accelerate hydrogen release of AB polymer complex, polyaminoborane (PAB). Polyaminoborane (PAB) is the spent fuel of ammonia borane (AB, NH_3BH_3) thermolysis with around 1 mole hydrogen release per mol AB. Dehydrogenation of PAB by thermolysis generally commences at temperatures above 120 °C. However, we find gaseous CO_2 environments dramatically enhance the thermolytic dehydrogenation of polyaminoborane (PAB) at 85 - 90 °C. By exposing PAB to 2.07 MPa CO_2, 5.48 wt % hydrogen based on the mass of ammonia borane (AB) desorbs in less than 1 h, while 4 h is required to generate 5.04 wt % hydrogen by maintaining PAB at 120 °C without using CO_2. The total hydrogen desorbed from the initial AB is 11.38 wt % under the CO_2 atmosphere at 85 °C, including the hydrogen release during the transition from AB to PAB. The main contributing factors to the facile hydrogen release are the promoting effect of newly formed formic acid and the exothermic effect of the reaction between PAB and CO_2. We will explain the possible promoting mechanism by various spectroscopic and elemental analyses.

This page intentionally left blank.

(134j) Synthesis of Nanoporous Palladium Powder With Controlled Pore and Particle Size for Hydrogen Storage Applications

Cappillino, P., Sandia National Laboratories
Jones, C.,
Hattar, K., Sandia National Laboratories
Clark, B.,
Hekmaty, M.,
Jacobs, B.,
Robinson, D. B., Sandia National Laboratories
Originally presented on: 11/4/2013 14:45:00 - 15:00:00

Synthesis of Nanoporous Palladium Powder with Controlled Pore and Particle Size for Hydrogen Storage Applications

by Patrick J. Cappillino, Christopher G. Jones, Khalid M. Hattar, Blythe G. Clark, Michelle A. Hekmaty, Benjamin W. Jacobs, and David B. Robinson

Materials capable of rapidly storing and delivering hydrogen are currently in demand for numerous energy applications.Â The high surface area exhibited by nanoporous palladium has the potential to greatly improve its kinetics of hydrogen charge and discharge *versus* nonporous metal while retaining favorable operating temperatures, pressures, and volumetric capacity.Â In addition, the void space imparted by porosity should accommodate the volumetric expansion of palladium upon hydriding, mitigating plastic deformation and improving cycle life.

By chemically reducing palladium salts in the presence of various surfactants, we are able to synthesize palladium having size-tunable mesopores from a few to tens of nm, in a scalable fashion.Â Our results suggest that Pd nanoparticles form and sinter around micelles present in the aqueous media.Â Changing the chemical composition of the surfactant affects the size of these micelles and ultimately determines the pore dimensions.Â We demonstrate that pore size affects not only surface area, but also pore thermal stability under vacuum, and in the presence of hydrogen gas.

In addition to pore geometry, particle size and shape are important factors in determining gas flow characteristics. Using nonporous copper particles as sacrificial reductants for palladium salts, in the presence of surfactants, we are able to synthesize micron-sized particles of uniform size and shape.Â This is in contrast to continuous aggregates and films that are obtained using soluble chemical reductants or planar electrodes.

We will present details of the synthesis of nanoporous palladium, demonstrating control of both pore and particle size.Â Bulk measurements of hydrogen storage properties and pore characteristics, as well as microscopic measurements, including *in situ* heated-stage TEM under vacuum and in the presence of hydrogen, will be included, as well as details on the kinetics of hydriding/dehydriding.

Sandia National Laboratories is a multi-program laboratory managed and operated by Sandia Corporation, a wholly owned subsidiary of Lockheed Martin Corporation, for the U.S. Department of Energy's National Nuclear Security Administration under contract DE-AC04-94AL85000.

(192a) Fabrication of Porous Carbon Nanofibers With Adjustable Pore Sizes As Electrodes for Supercapacitors

Tran, C., Drexel University
Kalra, V., Drexel University
Originally presented on: 11/4/2013 15:15:00 - 15:45:00

We report a facile method for obtaining extremely high surface area and uniformly porous carbon nanofibers for supercapacitors. Blends of polyacrylonitrile and a sacrificial polymer at different compositions have been electrospun into non-woven nanofiber mats with diameters in the range of 200–400 nm. Electrospun nanofiber mats are then subjected to carbonization to obtain porous carbon nanofibers (CNFs) as polyacrylonitrile converts to carbon and the sacrificial polymer decomposes out creating intra-fiber pores. Resultant porous CNFs exhibit specific surface area of up to 1600 $m^2 g^{-1}$ with a large fraction of mesopores (2–4 nm). No additional chemical or physical activation process was used. We demonstrate the tunability of the pore sizes within CNFs by varying the composition. The non-woven fiber mats of porous CNFs are studied as free-standing electrode materials for supercapacitors eliminating the need for polymeric binding agents. Electrochemical measurements showed large specific gravimetric and volumetric capacitances of up to 210 $F g^{-1}$ and 60 $F cm^{-3}$ in 1 M H_2SO_4 at a high cyclic voltammetry scan rate of 100 $mV s^{-1}$ due to the large fraction of mesopores. These materials retain 75% performance at a large current density of 20 $A g^{-1}$ indicating excellent power handling capability.

Nanomaterials for Energy Applications

(192b) Enhanced Energy Storage By Tunable Electrolyte Confinement in Structure-Directed CNT Arrays

Nicol, K. T., Mainstream Engineering Corporation

Zastrow, D., Mainstream Engineering Corporation

Hill, J. J., Mainstream Engineering Corporation

Originally presented on: 11/4/2013 15:45:00 - 16:00:00

The demand for power and energy dense storage devices is growing rapidly. While electrochemical double layer capacitors (EDLCs) offer high power density, many believe that high energy density is also obtainable through precise tuning of the electrolyte, electrode and structure of the overall cell. Multiwalled carbon nanotubes (MWNTs) are a promising electrode material for EDLCs because of their high electrical conductivity, chemical stability and exceptionally high surface area. Until single walled carbon nanotubes can be produced as solely conductive structures, dense and low-diameter MWNTs offer the best properties with regard to carbon-based EDLC electrode materials. When fabricated in anodic aluminum oxide templates, a MWNT is grown within each nanopore, resulting in an ultra-high density aligned array. This method produces a carbon electrode material with a significantly increased active surface area over commercially available materials. The energy density and capacitance of these materials can be further improved by tuning the size of the electrolyte domain around each MWNT. This is accomplished through partial and selective removal of the template, thereby confining the electrolyte domain. In typical EDLCs, the diffuse part of the double layer does not contribute to charge storage. However, we have observed an apparent charge storage enhancement in domain-confined MWNT electrodes. We will also discuss the tunable approach with regard to optimizing energy storage independent of the choice of REDOX couple or solvent.

(192c) Multifunctional Nitrogen-Rich â€œBrick-and-Mortarâ€ Carbon As High Performance Supercapacitor Electrodes and Oxygen Reduction Electrocatalysts

Yu, D.,
Wei, L., Nanyang Technological University
Chen, Y., Nanyang Technological University
Originally presented on: 11/4/2013 16:00:00 - 16:15:00

Novel porous carbon materials with excellent electronic, chemical and structural properties and high nitrogen content are desirable for many applications. Here, we show the design and synthesis of a new multifunctional porous carbon material with unique "point to plane" architecture through a simple but effective activation-free procedure. Porous carbon derived from a new type of copolymer poly (vinylidene chloride-co-acrylonitrile) serves as the nitrogen-rich "mortar". Reduced graphene oxide layers work as "bricks" with an aim to provide an open nanoscale scaffold and connect porous carbon, as well as modulate the ratio between mesopores and micropores. This new material has large surface area (957 m^2/g), high nitrogen content (6.6 at %), excellent conductivity (up to 5.1 S/cm), and favorable hierarchical meso- and microporosity. Benefiting from these intriguing features, this material shows ultrahigh specific capacitance at 361 F/g in aqueous electrolyte, far superior to that of many carbonaceous materials reported so far. The as-assembled asymmetric supercapacitors with the designed carbon material as negative electrodes and porous cobalt oxide nanorods as positive electrodes deliver a high-energy of 50.1 Wh/kg with a cell voltage of up to 1.6 V. Further, this material shows a superb electrocatalytic activity for oxygen reduction reaction in alkaline medium comparable to that of 20 wt% platinum-carbon electrodes, with much better methanol tolerant and long-term durability. We expect that this uniquely designed nitrogen-rich porous carbon material with its simple and scalable synthesis method will have great potentials for various applications in energy storage, energy conversion and catalysis.

(192d) Multiwalled Carbon Nanotubes With Tuned Surface Functionalities for Electrochemical Energy Storage

Wei, H., Lamar University
Gu, H., Lamar University
Guo, J., Lamar University
Wei, S., Lamar University
Guo, Z., Lamar University
Originally presented on: 11/4/2013 16:15:00 - 16:30:00

In the paper, surface functionalities effect on the electrochemical properties and energy storage properties of multiwalled nanotubes (MWNTs) were investigated via two different functionalized MWNTs, one modified with ether and carboxylic acid functional groups (denoted as MWNTs-EC), and the other with ester functional groups (denoted as MWNTs-E) as the supercapacitor materials. The functionalities onto the nanotubes were confirmed by Raman spectra and caused a minor enhancement in the band gap obtained from the Uv-vis spectra. Cyclic voltammetry results reveal that the capacitance of the MWNTs-EC, and MWNTs-E, was significantly increased by 41.5% and 123.2%, respectively, when compared to the as received MWNTs at lower scan rates. The energy density is also found to increase by 36.1 and 130.0% for the former and the latter, respectively, than that of the as received MWNTs-As in the galvanostatic charge/discharge measurement. The increases in both the capacitance and the energy density might arise from the enhanced surface wettability and larger pseudocapacitance contributions as confirmed by the electrochemical impedance spectroscopy. An equivalent circuit was proposed to disclose the nature of the capacitive behavior of the electrodes. The cyclic stability studies show an excellent long term cycling stability of the MWNTs, which retain 100.0% of the charge storage or discharge capacity even after 1000 charge-discharge cycles.

(192e) Three-Dimensional Core@Shell Nanostructured Array for Microscale Electrochemical Energy Storage

Mao, Y., Washington State University
Sun, X., Kansas State University
Originally presented on: 11/4/2013 16:30:00 - 16:45:00

Microelectromechanical systems (MEMS) based autonomous and biomedical devices play an ever increasing role in our society by monitoring the structural health of major civil works, the environmental presence of toxic or explosive compounds, and the physical health of human patients. Integration of robust and high power energy storage devices within these systems is important for reliable MEMS with necessary burst energy for data relay and dense components by reducing bulky external power sources. To overcome the shortcomings of conventional two-dimensional battery and supercapacitor configurations while developing more powerful energy storage devices, we focus on the development of three-dimensional architectures to take advantage of the vertical dimension-height. In this study, we designed a novel architecture composed of branched nanowire arrays, i.e. forest of nanotrees, to further boost electrochemical performance over one-dimensional nanowire arrays, inspired by branched trees with larger surface area to capture more sunlight. The prepared three-dimensional ZnO@MnO_2 core@shell branched nanowire arrays exhibit five times higher areal capacitance, better rate performance and smaller inner resistance than their nanowire array counterparts. These novel three-dimensional architectures offer promising designs for powering microelectronics and other autonomous devices on exceptionally small geometric scales.

(192f) Microfabricated Nickel Oxide Supercapacitors Based On High Aspect Ratio Concentric Cylindrical Electrodes

Armutlulu, A., Georgia Institute of Technology
Bidstrup Allen, S. A., Georgia Institute of Technology
Allen, M. G., Georgia Institute of Technology
Originally presented on: 11/4/2013 16:45:00 - 17:00:00

Supercapacitors are known as high-power energy storage devices, which narrow the application gap between the batteries and traditional electrostatic capacitors. As in the case of their electrostatic counterparts, they possess rapid charge and discharge capabilities with high degree of reversibility. Yet they can also store more energy per unit mass and/or volume which makes them superior to traditional capacitors and thus, yields a more attractive candidate for energy storage systems to be utilized in portable electronic devices. Supercapacitors can store energy electrostatically via separated charges in a Helmholtz double-layer that forms at the electrode/electrolyte interface, as well as electrochemically based on a phenomenon called pseudocapacitance that relies on reversible Faradaic charge transfer between the electrolyte and the surface of the electrode. In both energy storage mechanisms, electrode surface area plays a crucial role. In this study, our focus is on MEMS-enabled nickel oxide (NiO) based supercapacitors comprising high surface area 3D microelectrodes. The fabrication process consists of four major steps: formation of the array of high aspect ratio SU-8 pillars via backside exposure, deposition of alternating Cu and Ni layers on the SU-8 pillars by a sequential robotic electroplating process, mechanical and chemical etching of sacrificial Cu layers to obtain high surface area Ni structures in the form of concentric cylinders, and lastly conformal electrodeposition of the active material, i.e. nickel hydroxide ($Ni(OH)_2$), on the high surface area Ni backbone. Following the electrodeposition process of the $Ni(OH)_2$ layer, the structure is heated up to 300 ^oC in order to convert $Ni(OH)_2$ film to NiO, and the fabrication process is finalized. Performance tests for the resulting structures are carried out via cyclic voltammetry in 1 M KOH solution. Devices are cycled in excess of 500 times, and no significant capacitance degradation is observed. The areal and specific capacitance values of the supercapacitors are estimated to be 270 mF/cm^2 and 56 F/g, respectively.

(192g) Pretreatment Effects On Charge Storage of Early Transition-Metal Carbides and Nitrides

Djire, A., University of Michigan
Pande, P., University of Michigan
Sleightholme, A. E. S., University of Michigan
Deb, A., University of Michigan
Rasmussen, P. G., University of Michigan
Penner-Hahn, J., University of Michigan
Thompson, L. T., University of Michigan
Originally presented on: 11/4/2013 17:00:00 - 17:15:00

Pretreatment Effects on Charge Storage of Early Transition-Metal Carbides and Nitrides

Abdoulaye Djire[a], Priyanka Pande[a], Aniruddha Deb[b], Alice E. S. Sleightholme[a], Paul Rasmussen[ac], James Penner-Hahn[b], Levi T Thompson[ac]*

[a] Department of Chemical Engineering

[b] Department of Chemistry

[c] Hydrogen Energy Technology Laboratory

University of Michigan, Ann Arbor, MI_- 48109-2130

E-mail: ltt@umich.edu

Early transition-metal carbides and nitrides are promising candidates for use in supercapacitor electrodes due to their high electronic conductivities, high surface areas (can exceed 200 m^2/g), good electrochemical stabilities and high capacitance [1,2]. For example, the capacitance for VN has been reported to be as high as 1340 Fg^{-1} in aqueous KOH [3]. This high capacitance has been attributed to pseudocapacitive charge storage involving fast near-surface redox reactions. We observed that pretreatment of the surface to remove the passivation layer significantly improved the capacitances and electrochemical stabilities of early transition-metal carbides and nitrides in aqueous electrolytes. In this paper we present results from electrochemical characterization including cyclic voltammetry (CV), chronopotentiometry, and electrochemical impedance spectroscopy (EIS), and surface characterization using x-ray photoelectron spectroscopy (XPS) to explain changes in surface chemistry and electrochemical properties caused by pretreatment.Â

The early transition-metal carbides and nitrides of V, W and Mo were synthesized via temperature-programmed-reaction of their oxide precursors with 15% CH_4/H_2 or anhydrous NH_3 followed by passivation in 1% O_2/He at room temperature to form a oxygen-rich passivation layer preventing bulk oxidation on exposure to air [1]. Characterization of the structural properties was performed using nitrogen physisorption (BET surface area) and X-ray diffraction. Prior to the electrochemical characterization, the electrodes were pretreated in 0.3M NaOH aqueous solution for 1 min then rinsed with acetone, ethanol, and ultrapure water [4]. The CV was used to establish the stability windows and capacitances for these materials. The capacitance was deconvoluted into double-layer capacitance and pseudocapacitance by means of CV and EIS. Chronopotentiometry was used for further stability measurements and to determine changes in surface activities caused by pretreatment.

As shown in Figure 1, the specific capacitance for VN increased by 57% after pretreatment. This enhancement in performance suggests significant improvement in material electrochemical activities. We believe that pretreatment altered the charge storage for the carbides and nitrides. We also observed good electrochemical stability after successive cycling of the material for several thousand cycles. In general, the treated materials were more stable than the passivated materials.

Figure 2 shows the open circuit potential (OCP) or rest potentials for the passivated and treated VN in acidic medium. The OCP for the treated material is lower than that for the passivated material. We attributed this to changes in surface activities due to pretreatment.Â Using the Nernst equation, we found that the treated VN had a higher pseudocapacitance than the passivated VN. This pseudocapacitive charge-storage mechanism is being investigated using XPS and *in-situ* x-ray absorption spectroscopy.

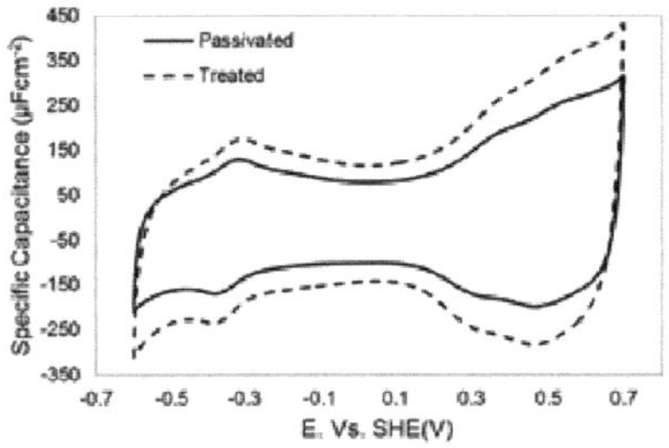

Figure 1: CV of passivated and treated VN in 0.1M H_2SO_4 at 50 mVs^{-1}.

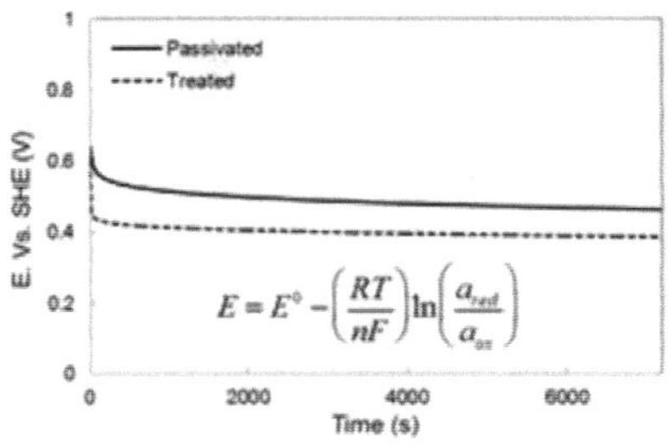

Figure 2: Rest potential of passivated and treated VN in 0.1M H_2SO_4

The physical and electrochemical properties will be described in terms of a proposed mechanism for psuedocapacitive charge storage.

References

(1) Cladridge, J. B.; York, A. P. E.; Brungs, A. J.; Green Malcolm L. H.; *Chem. Mater.* **2000**, 12, 132.

(2) Wixom, M.R.; Tarnowski, D. J.; Parker, J. M.; Lee , J.Q.; Chen, P. ?L.; Song, I.; Thompson, L. T.; *Mat. Res. Soc. Symp. Proc.* **1998,** 496, 643.

(3) Choi, D.;Â Kumta, P. N.; *Electrochem. Solid-State Lett.* **2005,** 8, 8, A418.

(4) Weidman, M.C.; Elposito, D.V.; Hsu, I. J.; Chen J. G.; J. Electrochem Soc.**2010**, 157, 12, F179

(192h) Doped Transition Metal Oxide Composite Electrodes for Supercapacitor Applications

Jampani Hanumantha, P., University of Pittsburgh
Kadakia, K., University of Pittsburgh
Hong, D. H., University of Pittsburgh
Poston, J., National Energy Technology Laboratory, U.S. Department of Energy
Ayyakkannu, M., National Energy Technology Laboratory
Kumta, P. N., University of Pittsburgh
Originally presented on: 11/4/2013 17:15:00 - 17:30:00

There is a dire need to provide a sustainable solution to the growing global demand for energy. It is this incessant demand that is the driving force fueling various current and emerging technologies including lithium ion batteries, supercapacitors and fuel cells. Supercapacitors occupy a unique place among these energy storage devices on account of their high power density and long cyclability. Supercapacitors are electrochemical charge storage devices capable of delivering large bursts of charge over a very short period of time. It is due to this unique ability that supercapacitors assume a very important role in the growing area of alternative energy storage systems. High surface area carbons, hydrated ruthenium oxide and MnO_2 are among the primary materials being considered for supercapacitor applications. Bulk oxide materials are traditionally hamstrung on account of their limited electrical conductivity for various electronic and electrochemical applications. In order to gain a fundamental understanding of the charge storage mechanisms in oxide materials, we deposited films of undoped transition metal oxides and doped transition metal oxides on current collectors and used various materials and electrochemical characterization techniques including X-ray photo-electron spectroscopy (XPS), cyclic voltammetry and electrochemical impedance spectroscopy to obtain a first-hand understanding of the underlying mechanisms. Effect of coating the various oxides on vertically aligned carbon nanotubes was also examined. Results of these studies will be presented and discussed.

Nanomaterials for Energy Applications

(192i) Novel 3-D MnO2/Holey Graphene Nanostructure for Supercapacitor Applications With Enhanced Electrochemical Performances

Wang, L., Wayne State University

Meng, T., Wayne State University

Ara, M., Wayne State University

Deng, D., Wayne State University

Ng, S., Wayne State University

Originally presented on: 11/4/2013 17:30:00 - 17:45:00

As an excellent energy storage device with high-power density and long life cycle, supercapacitor has been intensely investigated in many aspects, such as electrode materials, electrode structure engineering, electrolyte compositions, separator improvements, as well as device design. Among them, one of the most urgent shortcuts is the around 10 times lower energy density compared to the traditional Li-ion batteries. This drawback has dramatically confined the applications of supercapacitors in electrical vehicles, intermittent energy storages and other occasions where both the power density and energy density are required. However, many researchers have been working on improving the energy density without significantly scarifying power density. These researches include increasing the voltage windows by employing organic electrolyte with larger voltage openings, applications of thinner and yet still mechanically strong separators, integrate materials with higher intrinsic capacitance, and so on. Despite all the effort on supercapacitor research, the current state-of-art on energy density still needs to be improved to meet the fast-growing application demanding in the emerging and potential applications. Therefore, we propose a mild method of preparing a novel 3-D MnO2/holey graphene composite for supercapacitor electrodes. The nano MnO2 nanorods in the composites serves as a capacitance booster due to its high intrinct capacitance and spacer to prevent 2-D graphene sheets from stacking back together. At the same time, graphene also serves both as active material to provide large surface area for charge storage and conductive agent to collect electrons generated by the surface redox reactions of semiconductive MnO2. Holes on graphene sheets enhance the efficient accessibility of electrolyte to the surface of active materials. With this novel 3-D composite, an increasement of around 30% to 50% in capacitance is expected.

(192j) Electrochemical Energy Storage of Magnetic Carbon Nanocomposites: Role of Magnetocapacitance and Magnetohydrodynamics

Zhu, J., Lamar University
Wei, S., Lamar University
Guo, Z., Lamar University
Originally presented on: 11/4/2013 17:45:00 - 18:00:00

Magnetic field has been used in different chemical processes to improve the quality of electrodeposited films, yield of photoredox products, and oxidative reactivity of catalysts and *etc*. In this work, a small external magnetic field will be imposed on a supercapacitor unit trying to improve its electrochemical energy density. Magnetic microtubular carbon nanocomposite fabrics decorated with uniformly distributed metal oxides (iron, cobalt and nickel oxides) nanoparticles will be fabricated via both microwave-assisted and conventional annealing processes. These fabrics could serve as flexible electrodes for supercapacitors to deliver high energy density and exhibit excellent cycling stability. The effect of metal oxide species, annealing method and applied magnetic field on the capacitance of each nanocomposite material will be tested using cyclic voltammetry and galvanostatic charge/discharge measurements. The corresponding mechanism will be investigated using electrochemical impedance spectroscopy. Cycling stability of each composite with and without magnetic field will be also comparatively studied.

(192k) Synthesis and Characterization of Orange Peel As Electrodes in Li-Ion Capacitors

Arie, A. A., Parahyangan University

Originally presented on: 11/4/2013 18:00:00 - 18:15:00

Orange peel was utilized as the raw material for activated carbon-based electrodes in Li-ion capacitors. The carbon samples weres synthesized
using a chemical activation process followed by a surface modification method. The surface modification method showed no significant effect on the specific surface area, however it changed the surface property of the carbon samples. Experimental results demonstrated that the electrochemical characteristics of un-treated activated carbons had a better specific capacitance compared with those obtained by surface modification method. Based on our results, orange peel waste can potentially be used as a cheap raw material for the mass production of activated carbons as electrodes in Li-ion capacitors.

(196a) Microfluidic Droplet Dehydration For Separation and Purification Of Biomolecules

Anna, S. L., Carnegie Mellon University

Originally presented on: 11/4/2013 15:15:00 - 15:45:00

Droplets in microfluidic devices have proven useful as picoliter reactors for biochemical processing operations such as polymerase chain reaction, protein crystallization, and the study of enzyme kinetics. Although droplets are typically considered to be self-contained, constant volume reactors, there can be significant transport between the dispersed and continuous phases depending on solubility and other factors. In the present talk, we show that water droplets trapped within a microfluidic device for tens of hours slowly dehydrate, concentrating the contents encapsulated within. We use this slow dehydration along with control of the initial droplet composition to influence gellation, crystallization, and phase separation processes. For example, we show that a common aqueous two-phase system consisting of a PEG-Dextran mixture can be used to generate initially homogeneous droplets. Upon dehydration, the concentrated mixture phase separates, first into a microemulsion of Dextran droplets within the PEG phase, and later coalescing into two distinct droplets of nearly pure PEG and Dextran. The interfacial tensions of each phase with the continuous oil phase promote complete de-wetting of the two droplets, such that the PEG and Dextran phases separate into distinct, isolated droplets. Since DNA and proteins preferentially partition into PEG or Dextran, this phenomenon offers a useful method for on-chip separation and purification of samples containing these important biomolecules.

2013 Annual Meeting of the American Electrophoresis Society (AES)

(196b) Electrodes for Microfluidic Control and Sensing

Originally presented on: 11/4/2013 15:45:00 - 16:15:00

This presentation will cover microfabricated electrodes for control and sensing in microfluidic systems. The electrodes here are chemically inert and are used to drive dielectrophoretic flows or to sense impedance changes due to flowing particles. For some of these applications, submicron thickness “thin-film” electrodes are good enough. Thin-film platinum electrodes work for impedance sensing, and are easily integrated into planar fabrication during construction of polymer microfluidics or glass-based microfluidic systems. Besides sensing, these electrodes can also perform electrolysis to inject bubbles and ions into long-lived microfluidic devices if the materials are chosen carefully. For more diverse applications, thick electrodes (>1 micron) are important. A very common motivation for thick electrodes is to create a uniform electric field on opposite sides of a channel for sensing impedance changes caused by particles. Another motivation is to control the direction of and increase the volume occupied by fields in electrically-driven microfluidic flows. However, thick-film fabrication is more costly than thin-film, typically requiring a slow electroplating step that happens outside the planar fabrication sequence. This problem has driven the development of many workarounds to integrate thick electrodes in the microfluidic chip fabrication process. Conductive liquid electrode channels work well when the liquid electrode need not make direct contact with the fluid, for example when driving AC dielectrophoretic flows or doing capacitive sensing. For resistive sensing, ion injection, or performing induced-charge electroosmosis on metal surfaces with different spatial orientations, though, direct contact with thick electrodes is necessary, creating a fabrication challenge. A useful question is whether the thick electrodes really need to be solid in a given application, or whether they can be a thin film over a three-dimensional insulating shape. The presentation will include three alternate methods for patterning thin film over topography or for creating three-dimensional objects from released thin films. These techniques do not require electroplating and can be applied when it is only the three-dimensional electrode shape that matters. Examples will be presented from induced-charge electroosmosis.

(196c) Merging Electrical and Centripetal Forces With An Enzyme Cascade On a Compact Disc for the Ultimate in Analytical Performance in Molecular Diagnostics

Madou, M. J., University of California, Irvine
Kamath, R.,
Peytavi, R.,
Originally presented on: 11/4/2013 16:15:00 - 16:45:00

In this contribution we introduce the merging of three platform technologies to achieve optimal lower limits of detection at a low cost for point of care molecular diagnostics. The first platform technology entails inexpensive Carbon-MEMS based Interdigitated Electrode Arrays (IDEAs) that achieve a redox amplification factor of more than 30. Detection limits of attomols are within reach with this technology. The 3D carbon IDEAs are fabricated by first nano-imprinting 3D patterns on a polymer precursor followed by pyrolysis of the polymer in an inert atmosphere. The second platform technology is based on specific DNA detection using an enzymatic cascade of a set of kinetically well-balanced enzymes: DNA polymerase, ATP sulfurylase, flavin adenine dinucleotide (FAD) synthetase and an apo-enzyme (apo-glucose oxidase). PPi generated by isothermal DNA amplification is converted into ATP by the ATP sulfurylase that is used by the FAD synthetase to produce FAD. FAD diffuses to an enzymatic electrode where apo-glucose oxidase has been entrapped in a hydrogel. Apo-glucose oxidase is a glucose oxidase without its prosthetic factor FAD. In the presence of FAD produced by the cascade, the reconstituted glucose oxidase, produces thousand of electron per second in presence of glucose. The third platform is the CD microfluidic platform, that not only automates and multiplexes the enzyme cascade assay but further increases the lower limit of detection because of the flow of the analyte over the redox amplifying electrodes. In the case demonstrated a reciprocating system fluidic system on the CD, using a micro-balloon system, allows for flow measurements at relatively low rpm and with very small amounts of sample.

(196d) Electrokinetics and High Pressure Liquid Chromatgraphy

Arnold, D., Eksigent

Originally presented on: 11/4/2013 16:45:00 - 17:15:00

Fundamental investigations of electrokinetic flow led to the founding of Eksigent. From this starting point, widely varying products and applications for the origen have taken Ekisgent into high pressure liquid chromatography, chip based systems, and mass spectrometry solutions. These investigations have led to an expertise in precision flow control and fluid interfaces. Applications of these innovations range from drug discover, biomarker discovery and measurement to proteomics, genomics, and metabonomics.

Protein identification and functional analysis require new analytical tools and techniques that offer high throughput and enhanced detection sensitivity for identification of low-abundance proteins. To meet this need, Eksigent developed NanoLC™ systems with extremely precise gradient control at the 100 to 200 nL/min column flow rates preferred for nanospray MS, with integral peak parking. Multiplexing the system to allow multiple, simultaneous separations increases throughput substantially when used in combination with the cHiPLC®-Nanoflex System or with high-throughput MALDI systems, like the EKSpot™ MALDI Spotting System. Eksigent's ExpressLC® system has unique innovations to provide increased levels of chromatographic resolution and speed at low sample volumes and flow rates.or conducting library purity tests, developing validated methodologies for ADMETox studies or completing formulation and stability testing in the drug discovery and development industry.

(196e) Single Cell Genomics

Quake, S., Stanford University and Howard Hughes Medical Institute

Originally presented on: 11/4/2013 17:15:00 - 17:45:00

An exciting emerging area revolves around the use of microfluidic tools for single-cell genomic analysis. We have been using microfluidic devices for both gene expression analysis and for genome sequencing from single cells. In the case of gene expression analysis, it has become routine to analyze hundreds of genes per cell on hundreds to thousands of single cells per experiment. This has led to many new insights into the heterogeneity of cell populations in human tissues, especially in the areas of cancer and stem cell biology. These devices make it possible to perform “reverse tissue engineering” by dissecting complex tissues into their component cell populations, and they are also used to analyze rare cells such as circulating tumor cells or minor populations within a tissue. We have also used single-cell genome sequencing to analyze the genetic properties of microbes that cannot be grown in culture—the largest component of biological diversity on the planet—as well as to study the recombination potential of humans by characterizing the diversity of novel genomes found in the sperm of an individual. We expect that single cell genome sequencing will become a valuable tool in understanding genetic diversity in many different contexts.

(252a) Low-Cost 3D-Printed Electrokinetic Systems

Bengtsson, K., LinkÃ¶ping University
Robinson, N. D., LinkÃ¶ping University

Originally presented on: 11/5/2013 8:30:00 - 8:52:00

Automated 3D-printing, or fabbing, of microfluidic systems is revolutionizing the development of miniaturized chemical and biological lab-on-a-chip devices. Such systems can be rapidly produced, in low volumes, by subtractive (removal of material) or additive (addition of material) processes controlled by a computer. Of these types, additive fabrication (printing) offers the opportunity to use multiple materials, including even chemical reagents and biological samples, in the systems being produced. To date, nearly all of the 3D-printed microfluidic systems published are static or passive — that is to say that they require external pumps, magnetic fields, or other equipment to make them functional — usually because the various materials required for electrodes and other active functions are difficult to incorporate into a single fabrication process.

Capitalizing on recent advancements in (2D) printed materials for organic- and bio-electronics, we demonstrate bottom-up 3D-printing of functional lab-on-a-chip devices employing conventional capillary electrophoresis. These systems contain printed electrodes and fluidic channels, and require only an external power source (which could alternatively be printed in the chip). By employing electrochemically-active conjugated polymer electrodes, we can readily print our devices straight from a syringe-based extrusion fabber, and simultaneously reduce the influence of electrolytically-generated by-products during sample transport and separation. [1]

Our intention is to build on our experience with these devices and improve the materials, methods and device structures available to make local on-demand production of point-of-care medical diagnostics possible.

1. Erlandsson, P. G.; Robinson, N. D., Electrolysis-reducing electrodes for electrokinetic devices. *Electrophoresis* 2011, **32**, 784-790.

(252b) Inexpensive Electrokinetic Microfluidic Systems Using Printed Circuit Board Substrates

Romero, N.,
Parkes, L.,
Jackson, D.,
Naber, J.,
Williams, S. J., University of Louisville
Originally presented on: 11/5/2013 8:52:00 - 9:14:00

Dielectrophoresis (DEP) is a well-known particle manipulation technique that has been able to capture, sort, concentrate, and characterize a variety of biological and artificial particles down to nanometers in diameter. Within the past decade there has been a thrust in developing novel DEP microsystems with the goals of higher throughput, inexpensive fabrication, and user-friendliness. Recent DEP techniques have developed fabrication methods to incorporate 3D electrodes or structures for improved throughput; however, clean room facilities and equipment are needed in order to create these platforms, limiting the number of researchers and students exposed to such technology.

This work shows preliminary results of fabricating 3D DEP microfluidic systems using neither clean room facilities nor photolithography. Rather, fabrication uses a CNC milling machine equipped with a sub-millimeter end mill. A copper-clad board provides the substrate for the system. Milled features serve as the microfluidic network with the copper electrodes being the sidewalls of the channels. The procedure of milling copper-clad boards is well known by the printed circuit board (PCB) industry and hobbyists. In addition, sub-millimeter end mills are commercially available for under $50 including 25, 50, and 100 micrometers. Preliminary results demonstrate this concept, aggregating 8 micrometer beads at 10 V using 3D quadropole electrodes milled with a 100 micrometer bit.

(252d) Effect of Insulating Post Shape On Particle Trapping With Insulator-Based Dielectrophoresis

Gencoglu, A., Rochester Institute of Technology
La Londe, A., Rochester Institute of Technology
Koppula, K. S., Rochester Institute of Technology
Romero-Creel, M., Rochester Institute of Technology
Lapizco-Encinas, B. H., Rochester Institute of Technology
Originally presented on: 11/5/2013 9:14:00 - 9:36:00

Microfluidics is a rapidly growing field, electrokinetic techniques such as electrophoresis, electroosmosis and dielectrophoresis are becoming increasingly important due to their great flexibility. Dielectrophoresis (DEP) refers to particle movement due to polarization effect when particles are exposed to non-uniform electric fields. In insulator-based dielectrophoresis (iDEP) non-uniform electric fields are created by employing 3-dimensional insulating posts between two electrodes. The effect of the separation distance between insulating posts on the dielectrophoretic force force acting on particles and the effectiveness of iDEP techniques is well known. The present research studies the effect of the post geometry on the dielectrophoretic force and particle immobilization. Manipulations of inert polystyrene particles was carried out in microchannels containing arrays of insulating posts that had an effective diameter of 450 μm with a distance between the posts of 50 μm in all cases. Insulating post shapes including circle, diamond, half circle and triangle (half diamond) were compared. In addition to the experimental work, COMSOL software was used to model the electric field gradients and DEP force generated by each post geometry.

(252e) Frequency-Selective Polarization of the Electrical Double-Layer Around Nano-Colloids

Su, Y. H., University of Virginia
Varhue, W., University of Virginia
Tsegaye, M., University of Virginia
Rohani, A., University of Virginia
Swami, N., University of Virginia
Originally presented on: 11/5/2013 9:36:00 - 9:58:00

The direct observation of dielectrophoretic behavior of nanoparticles (sub-100 nm) due to Maxwell-Wagner interfacial polarization is limited by the dissipative action of electrothermal flow due to Joule heating, especially within media of moderate to high conductivity [1]. However, for nano-colloids of sizes comparable to their double layer thickness, the induced electrical dipoles are stronger than that predicted by Maxwell-Wagner theory due to significant ion migration in the normal direction to the applied field as a result of the substantial extent of the double-layer [2]. Prior work has demonstrated that the resulting ion adsorption onto the Stern region of the double-layer enhances its polarization and leads to a wider frequency spread for positive dielectrophoresis than observed for larger particles with thinner double-layers [3]. Herein, we study the influence of this ion migration normal to the field on the polarization behavior of the diffuse region of the double-layer around silica nano-colloids. Through utilizing successively smaller nano-colloids, we demonstrate that the net dipole can be reversed for particles below a critical size to cause negative dielectrophoresis, presumably due to the polarization of the diffuse region of the double-layer. Furthermore, we demonstrate that the frequency window for negative dielectrophoresis behavior can be enhanced through polarization of the nano-colloid within media of successively higher ionic strengths, to enhance ion migration. Since the conformation of biomolecules strongly determines the surface conductance and polarization behavior of the diffuse region of the double-layer [4], we envision that this phenomenon can be applied towards the frequency-selective electrokinetic separation of biomolecules, such as DNA aptamers, based on their molecular conformation state.

[1] V. Chaurey, A. Rohani, Y.-H. Su, W. Varhue, K.T. Liao, C. F. Chou, N. S. Swami, "Scaling down constriction-based dielectrophoresis for trapping nanoscale biomolecules in high conductivity media", *Electrophoresis* (2013), 34, 1097-1104.

[2] S. Basuray and H. -C. Chang, "Induced dipoles and dielectrophoresis of nanocolloids in electrolytes," *Phys. Rev. E,* vol. 75, p. 060501 (R), 2007.

[3] I. Ermolina and H. Morgan, "The electrokinetic properties of latex particles: comparison of electrophoresis and dielectrophoresis," *J of Colloid and Interfacial Science,* vol. 285, pp. 419-428, 2005.

[4] M. Hughes and H. Morgan, "Dielectrophoretic characterization and separation of antibody-coated submicrometer latex spheres," *Anal. Chem.,* vol. 71, no. 16, pp. 3441-3445, 1999.

(252f) Unsteady Electrohydrodynamic Drop Deformation

Lanauze, J., Carnegie Mellon University
Walker, L. M., Carnegie Mellon University
Khair, A. S., Carnegie Mellon University
Originally presented on: 11/5/2013 9:58:00 - 10:20:00

We quantify the transient deformation of a droplet immersed in a weakly conductive (leaky dielectric) fluid upon exposure to a uniform DC electric field. Capillary forces are assumed to be sufficiently large that the drop only slightly deviates from its equilibrium spherical shape. In particular, we account for transient (or linear) fluid inertia via the unsteady Stokes equations, and also account for a finite electrical relaxation time over which the drop interface charges. The temporal droplet deformation is governed by two dimensionless groups: (i) the ratio of capillary to momentum diffusion time scales: an Ohnesorge number *Oh*; and (ii) the ratio of charge relaxation to momentum diffusion time scales, which we denote by *Sa*. If charge and momentum relaxation occur quickly compared to interface deformation, $Sa \ll 1$ and $Oh \gg 1$ for the droplet and medium, a monotonic deformation is acquired. In contrast, $Sa > 1$ and $Oh < 1$ for either phase can lead to a non-monotonic development in the deformation. Numerical values for the deformation are calculated by inverting an analytical expression obtained in the Laplace domain, and are corroborated by asymptotic expansions at early and late times. The droplet and medium behave as perfect dielectrics at early times, which always favors an initial prolate (parallel to the applied field) deformation. As a consequence, for a final oblate (normal to the applied field) deformation, there is a shape transition from prolate to oblate at intermediate times. This transition is caused by the accumulation of sufficient charge at the interface to generate electrical and viscous shear stresses. Notably, after the transition, there may be an "overshoot" in the deformation, i.e., the magnitude exceeds its steady-state value, which is proceeded by an algebraic tail describing the arrival towards the final, steady deformation. Our work demonstrates that transient inertia or a non-zero electrical relaxation time can yield non-monotonic electrohydrodynamic drop deformation.

(252g) Hybrid-Field Microfluidics Enhanced Polyplex Synthesis and Delivery

Ren, F., Louisiana Tech University
Huang, S., Louisiana Tech University
Wang, S., Louisiana Tech University
Originally presented on: 11/5/2013 10:20:00 - 10:42:00

Micro/nanofluidics has great potentials for many biomedical applications, such as noviral gene delivery with polyplex and lipoplex. Hydrodynamic focusing microfluidics and other flow patterns have been successfully used for improving the synthesis quality of such nanoparticle probes. In these applications, it is cirtical to ensure efficient initial interactions between molecular probes and their later conjugation homogeneity. Therefore, it is highly desired to bring cationic polymers, DNA, or both to their conjugation interfaces for efficient packing and production of nanocarriers only overthere. Here we introduce a hybrid field microfluidic (HFM) approach to create such conjugation environment (interfacial conjugation) for efficient polyplex synthesis and delivery. In HFM, an electric field is imposed at desired locations and appropriate moments to balance the flow stress on negatively charged DNA molecules to effectively regulate their initial conformations, deformation dynamics, and accumulation in flows. Under appropriate conditions, DNA molecules could continuosly accumulate at the interface of different flow streams from the main flow in a hydrodynamic focusing setup, where the conjugation with polyethylenimine (PEI) occurs. Frequently switching the electric field helps retain most probes at the conjugation interface while avoid potential drawbacks such as large aggregates and Joule heating. Atomic force microscopy (AFM) and dynamic light scattering (DLS) results confirmed the good complexation quality of polyplex nnaoparticles. The delivery efficiency was evaluated with mammalian cells with better cellular uptake of molecular probes and reduced adverse effects. This new nanoparticle synthesis concept may stimulate further exploration in the delivery of various therapeutic materials for both *in vitro* and *in vivo* applications.

(301b) A Study of Doped Nonpolar Liquids Using Electrochemical Impedance Spectroscopy

Yezer, B., Carnegie Mellon University
Khair, A. S., Carnegie Mellon University
Sides, P., Carnegie Mellon University
Prieve, D. C., Carnegie Mellon University
Originally presented on: 11/5/2013 12:30:00 - 12:55:00

The electrical response of nonpolar media to an AC field is described and its conductivity, Debye length, and permittivity are determined.Â Surfactants are added to nonpolar media to increase the electrical conductivity and control particle dispersion in a number of industrial applications.Â Steric stabilization of dispersions is typically assumed in nonpolar media; electrostatic effects are less well characterized. Â Developments in this field require definitive measurements of electrical properties of the bulk fluid to determine mechanisms for the formation and stabilization of charges. Â Electrochemical impedance spectroscopy (EIS) was applied to a cell comprising parallel plate indium tin oxide electrodes separated by 10 µm of dodecane doped with a commercial surfactant OLOA 11000.Â At high frequency the charging of the electrodes was observed as a maximum in the out of phase, or imaginary, impedance response.Â Fitting a simple equivalent circuit of a resistor and capacitor in parallel to the high frequency response determined the electrical conductivity and permittivity of the fluid.Â At low frequency the charging of the electrical double layers at either electrode was observed as a minimum in the imaginary part of the impedance response and was fit with a 5 element equivalent circuit model to determine the Debye length of the fluid.Â The number of charge carriers in solution inferred from the Debye length measurement increased linearly with surfactant concentration.Â For example, the conductivity of 10% OLOA in dodecane was 1.7 Â± 0.1 nS/cm and the dielectric constant was 2.3 Â± 0.1.Â The diffusion coefficient of charge carriers calculated from the measured Debye length and conductivity was 25 Â± 3.6 $\mu m^2 s^{-1}$.Â A numerical solution of the Poisson-Nernst-Planck model that predicted the motion of charge carriers and current response inside the cell was in agreement with experimental observations at high and moderate frequencies.Â At frequencies below 10 Hertz, the real part of the cell impedance increased as frequency decreased, which might indicate unequal size of charge carriers. Â

(301c) Surfactant-Mediated Electrophoretic Properties of Non-Polar Dispersions

Lee, J., Georgia Institute of Technology
Behrens, S. H., Georgia Institute of Technology
Originally presented on: 11/5/2013 12:55:00 - 13:20:00

Charging in non-polar dispersions is important in a wide range of applications such as electrophoretic image displays and liquid toners. Compared to the well-known cases of aqueous media where electric charges are ubiquitous, charging in non-polar dispersions is much less common and intuitive because of the large energetic cost of generating or sustaining charges in such low dielectric media. Electric charging in non-polar liquids, however, is nonetheless observed, especially in the presence of surfactant additives. Surfactants are known to stabilize charges in the bulk liquid by incorporating ions in the core of reverse micelles and to promote the surface charging of particles dispersed in non-polar liquids. Despite its academic and industrial interest, the actual mechanism of particle charging in non-polar liquids is still poorly understood. Even though a number of hypothetical mechanisms have been proposed, these are largely system dependent, and most studies have either focused on a single surfactant type or compared a small number of vastly different systems, making the relative contribution of different charging mechanisms hard to identify. The objective of this study is to achieve a better understanding of the surfactant-mediated charging in non-polar dispersions via thorough characterization of the well-defined systems under systematic variation of the most relevant system parameters. In particular, we investigate the particle charging mediated by a series of custom-synthesized and well-purified polyisobutylene succinimide (PIBS) surfactants with small structural variations. The surface properties of charge-acquiring colloidal particles are also varied. Conductivity and electrophoretic mobility measurements assessing the charging effect of the surfactants are complemented by light scattering studies of the surfactant micelles, and the residual water content is assessed using Karl-Fischer titration. Acid-base interactions between the particle surface and the surfactant are studied quantitatively by interfacial tensiometry and contact angle measurements. Our experimental results indicate that smallest variations in surfactant chemistry result in an observable differences in the charging behavior, and that particle charging may indeed result from the interplay of several charging mechanisms.

Michor, E. L., University of Washington

Berg, J. C.,

Originally presented on: 11/5/2013 13:20:00 - 13:45:00

A growing number of applications seek to use electrostatic effects in apolar media for aggregation stability and manipulating particle movement in such systems, as exemplified by the Amazon Kindle® reader and HP ElectroInk® printer. Progress is limited, however, by incomplete knowledge of the mechanism(s) of particle charging in these systems. To help shed light on this, the current work focuses on the study of several common factors that affect the polarity and magnitude of particle charge in apolar systems: particle surface functional groups, surfactant functionality, trace water content, and the implementation of synergists.

One theory for the origin of particle charge in apolar media involves an acid-base charge transfer between the particle surface functional groups and adsorbed surfactant molecules. This theory is explored by measuring the electrophoretic mobility of a series of mineral oxide particles that are dispersed in Isopar-L using common charge-stabilizing surfactants: OLOA 11000 (basic), SPAN 80 (acidic), and AOT (ionic). The mineral oxide series is selected to yield a wide range of point of zero charge (pzc) values when in water (in terms of pH), which serves as an indication of the relative acidity or basicity of the particle surface functional groups. It is found that the polarity and magnitude of the particle charge is strongly dependent on the acid-base functionality of both the particle functional groups and the surfactants.

Another important factor that affects particle charging is the amount of trace water that is introduced to the system by hygroscopic particles and surfactants. The amount and location of this water can have significant effects on the electrical properties of these systems. The current work not only explores the effect of water content on particle charge in apolar dispersions, but differentiates between water bound to the particles and water located in inverse micelles. This is accomplished by measuring the electrophoretic mobility of silicon dioxide particles dispersed in solutions of Isopar-L and OLOA 11000. The water content is determined for both the dispersion and the supernate after centrifuging the particles out.

Finally, synergists are often employed to enhance the adsorption of surfactants onto particle surfaces, providing steric stabilization against aggregation. However, these synergists can also have a dramatic effect on particle charging in apolar systems by either adsorbing to the particle surface or becoming incorporated into the surfactant inverse micelles. In this study we examine the electrophoretic mobility of commercially available cyan and magenta pigments with their synergist counterparts dispersed in heptane with charge-stabilizing surfactants OLOA 11000, AOT, and Span 80.

(301e) Why Ionic Surfactants Exhibit Linear Conductivity-Concentration Dependence in Non-Polar Liquids

Dukhin, A., Dispersion Technology Inc
Originally presented on: 11/5/2013 13:45:00 - 14:10:00

It is known that addition of ionic surfactants to non-polar liquids enhances conductivity by orders of magnitude. Conductivity increases linearly with the concentration of the added surfactant. This fact was considered as contradicting to the classical dissociation theory and Bjerrum-Onsager-Fuoss theory of ion pairs in on-polar liquids. It serves as justification for introduction of alternative, so-called disproportionation theory.

We show that there is no contradiction between this fact and theory of ion-pairing. Mass action law of dissociation model predicts existence of a "critical ionic strength" below which ion-pairs do not form. Concentration of ions below this critical concentration is so low that rate of ions collisions cannot compensate for the rate of ion-pairs breakup. We have estimated this "critical ionic strength" using known size information of the inverse micelles ions formed by ionic surfactants. Indeed, it is orders of magnitude higher than ionic strength usually achieved with ionic surfactants.

In contrary, non-ionic surfactants do exhibit non-linear conductivity-concentration dependence. At the same time, non-ionic surfactants generate conductivity by orders of magnitude higher than ionic ones. Therefore they produce higher ionic strength that exceeds "critical ionic strength" of ion-pairs formation. Observed non-linearity is associated with ion-pairs formation in agreement with the classical theory.

(301a) Electrophoresis Of pH-Regulated Particles In The Presence Of Multiple Ionic Species

Tseng, S., Tamkang University
Hsu, J. P., National Taiwan University
Wang, N., National Taiwan University
Originally presented on: 11/5/2013 14:10:00 - 14:35:00

An electrophoresis model taking account of the pH-regulated nature of particles, and the presence of multiple ionic species is proposed for arbitrary surface potential and double layer thickness. The estimated zeta potential is compared with those estimated from conventional formulas usually adopted by conventional zeta potential instrumentation. Using Fe3O4 nanoparticles in NaCl solution with pH adjusted by HCl and NaOH as an example, we show that these formulas might yield appreciable deviations. For example, at pH 9, the deviation is one the order -25 mV. With the surface charge density measured by titration and the zeta potential by electrophoresis, the true surface potential is estimated through a triple-layer model.

Olson, D. W., University of Minnesota - Twin Cities

Joswiak, M.,

Park, S. G., University of Minnesota

Dorfman, K. D., University of Minnesota

Originally presented on: 11/5/2013 12:30:00 - 12:52:00

We have created a DNA filter that takes advantage of the nonlinear dependence of the electrophoretic mobility on applied electric field in the "nanofilter" array geometry to create a temporal asymmetry ratchet. The nanofilter array is a separation device that is made of a series of thin slits, typically 60 nm deep, which impedes DNA motion and leads to size-based separation of DNA [1]. However, it is not actually a filter. Instead, it acts as a gel-like sieving matrix with an electrophoretic mobility that depends nonlinearly on the electric field. It has been shown theoretically and in simulations that such devices can enact a ratchet-like behavior in DNA by periodically alternating between two different magnitude electric fields in opposite directions [2,3]. Running the nanofilter in this manner allows the device to act as a filter, only permitting DNA below a desired size through the array while not clogging like a traditional filter.

We will present experimental proof of the nanofilter array ratchet. Moreover, we will show that (i) the cutoff molecular weight for the filter can be adjusted by changing the electric field strength and/or pulse times without making a new device and (ii) the ratchet mode can be used to enhance the resolution of a separation. One could imagine collecting one set of the fragments then changing the electric fields to collect the second and so on to completely fractionate the mixture. This DNA filter is made using standard silicon microfabrication techniques, which lends itself to being integrated as a filtration step in a larger lab-on-a-chip type device for DNA processing.

REFERENCES:

1. J. Fu, P. Mao, and J. Han. *Appl. Phys. Lett.* **87**, 263902 (2005).
2. G. W. Slater, H. L. Guo, and G. I. Nixon. *Phys. Rev. Lett.* **78**, 1170 (1996).
3. F. Tessier and G. W. Slater. *Appl. Phys. A* **75**, 285 (2002)

See more Presentations from this session: ~~Nanoscale Electrokinetics~~

See more Sessions in this Group: ~~2013 Annual Meeting of the American Electrophoresis Society (AES)~~

(320b) Simultaneous Purification and Fractionation of Nucleic Acids and Proteins From Complex Samples Using Isotachophoresis

Qu, Y., Stanford University
Marshall, L. A., Stanford University
Santiago, J. G., Stanford University
Originally presented on: 11/5/2013 12:52:00 - 13:14:00

harp counter migrating bands. (Bottomng by ion by using

SIMULTANEOUS PURIFICATION AND FRACTIONATION OF NUCLEIC ACIDS AND PROTEINS FROM COMPLEX SAMPLES USING ISOTACHOPHORESIS

Yatian Qu, [1] Lewis A. Marshall, [2] Juan G. Santiago[1]

[1] Department of Mechanical Engineering, Stanford University

[2] Department of Chemical Engineering, Stanford University

Stanford, CA 94305, USA

We report on the development of a novel integrated sample preparation technique to simultaneously extract nucleic acids and proteins from complex biological samples using isotachophoresis (ITP). ITP is an electrophoretic technique that both separates and pre-concentrates ions based on their electrophoretic mobility. ADDIN EN.CITE <EndNote><Cite><Author>Persat</Author><Year>2009</Year><RecNum>10</RecNum><DisplayText><style Marshall, L.A., Santiago, J.G.</author></authors></contributors><titles><title>Purification of Nucleic Acids from Whole Blood Using Isotachophoresis</title><secondary-title>Analytical Chemistry</secondary-title></titles><periodical><full-title>Analytical Chemistry</full-title></periodical><pages>9507-9511</pages><volume>81</volume><number>22</number><dates><year>2009</year></dates><urls></urls></record></Cite></EndNote>[1] Our chip contains a single input sample reservoir connected to two channels leading to output reservoirs. Anionic ITP in one channel is used to extract and purify DNA and deliver it to the DNA output reservoir. Simultaneously, the second channel uses cationic ITP to extract and purify proteins, preserving their natural folding state. Purified DNA and proteins are thereby fractionated and delivered to respective output reservoirs. The protein extraction can be configured to exclude albumin, a highly abundant species with little analytical value. The DNA is purified away from species that inhibit enzymatic amplification such as polymerase chain reaction (PCR).

We designed and fabricated the novel microfluidic device shown in Figure 1 to demonstrate the technique. The device is fabricated from polydimethylsiloxane (PDMS) and glass and includes two "C" shaped channels leading from a single sample input reservoir at the center. The channels are 1 mm wide and 100 μm deep, with a volume of 4 μL for each branch. Each of two branches leading form the center is about 40 mm long. The cationic ITP channel collects positively charged protein species. The anionic ITP channel collects nucleic acids, which are strongly negatively charged. Each branch has its own elution reservoir, where the collected biomolecules can be pippetted off the chip. Each elution reservoir is also connected to a buffering channel and reservoir to provide additional pH-buffering capacity.

We successfully demonstrated simultaneous extraction of extracellular DNA and proteins from 1 μL human serum samples. We optimized the buffer chemistry for this system by performing numerical simulation using Stanford Public Release Electrophoretic Separation Solver (SPRESSO), as shown in Figure 2. We visualized the process in a small version of the separation device in order to obtain a series of images that capture the dynamics of the ITP sample zone in both branches, and show these images in Figure 3. For purification experiments, we imaged focused nucleic acids zone (Figure 4) by using fluorescent dye TOTO-3. The focused protein zone (Figure 5) was monitored by imaging spiked yellow fluorescent protein (YFP).

To verify the purification of nucleic acids, we performed quantitative polymerase chain reaction (qPCR) using primers for the human gene BRAC2 (Figure 6). Even though human serum contains low abundance of extra-cellular DNA, ADDIN EN.CITE <EndNote><Cite><Author>O'Driscoll</Author><Year>2007</Year><RecNum>8</RecNum><DisplayText><style Article">17</ref-type><contributors><authors><author>Lorraine O'Driscoll</author></authors></contributors><titles><title>Extracellular Nucleic Acids and their Potential as Diagnostic, Prognostic and Predictive Biomarkers</title><secondary-title>Anticanser Research</secondary-title></titles><periodical><full-title>Anticanser Research</full-title></periodical><pages>1257-1266</pages><number>27</number><dates><year>2007</year></dates><urls></urls></record></Cite></EndNote>[2] ITP successfully recovered sufficient material to detect this gene via qPCR. Unprocessed serum shows no amplification in qPCR, as expected due to the effects of PCR inhibitors present in serum such as immunoglobulin G (IgG). To analyze the extracted proteins, we performed sodium dodecyl sulfate polyacrylamide gel electrophoresis (SDS-PAGE) followed by silver staining. We compared samples extracted via ITP with samples directly applied to the gel, as shown in Figure 7. We recovered a wide variety of proteins from the original sample, and successfully excluded albumin.

Simultaneous extraction and purification of nucleic acids and proteins from a single crude biological sample is important for ensuring comparability between genomic and proteomic results, and to conserve precious samples. ADDIN EN.CITE <EndNote><Cite><Author>Butt</Author><Year>2007</Year><RecNum>3</RecNum><DisplayText><style Article">17</ref-type><contributors><authors><author>Butt, R. H., Pfeifer, T.A., Delaney, A., Grigliatt, T.A., Tetzlaff, W.G., Coorssen, J.R.</author></authors></contributors><titles><title>Enabling Coupled Quantitative Genomics and Proteomics Analyses from Rat Spinal Cord Samples</title><secondary-title>Molecular & Cellular Proteomics</secondary-title></titles><periodical><full-title>Molecular & Cellular Proteomics</full-title></periodical><pages>1574-1588</pages><number>6</number><dates><year>2007</year></dates><urls></urls><electronic-resource-num>10.1074/</electronic-resource-num></record></Cite></EndNote>[3] **To our knowledge, our technique is unique in offering availability of automated and simultaneous nucleic acids and protein extraction in microfluidics platforms.**

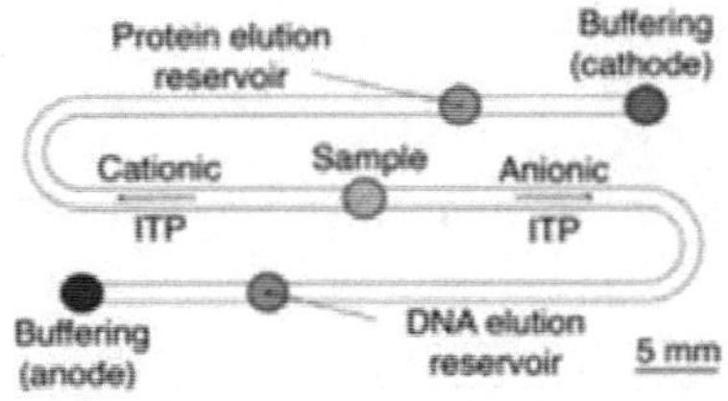

Figure 1. Device design for simultaneous purification of nucleic acids and proteins from serum using simultaneous cationic and anionic ITP processes. Each reservoir holds approximately 8 ÂµL liquid. Each separation channel has a volume of 4 ÂµL.

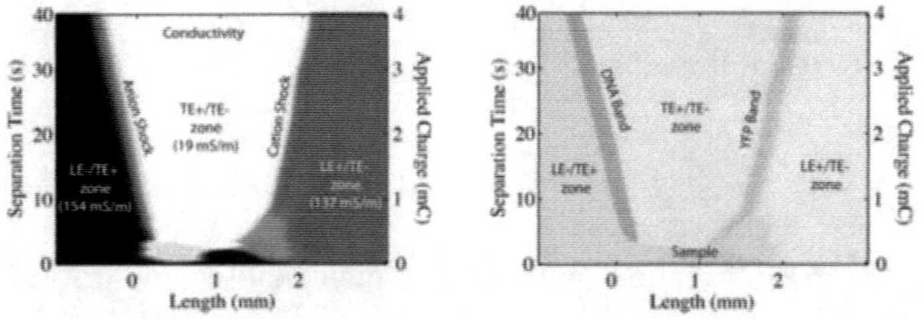

Figure 2. Spatiotemporal representation of numerical simulation results. (Top) Ion concentrations: each ion is given its own false color. Over time, the DNA and proteins focus into sharp counter migrating bands. (Bottom) Conductivity map: low conductivity causes high electric field in the TE zones; fast ions quickly migrate out.

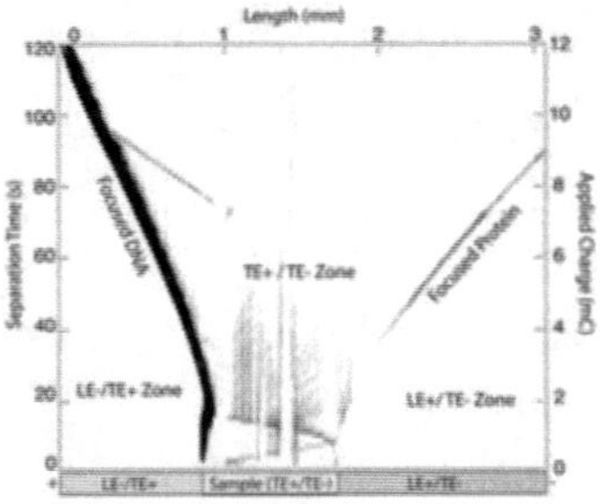

Figure 3. Experimental data plotted in a spatiotemporal plot showing the dynamics of protein and DNA zones purified in our chip. Yellow fluorescent protein and DNA labeled with SYBR Green I were imaged using a custom fluorescence set up using a stereoscope. The DNA and protein zones are clearly visible propagating away from the sample zone.

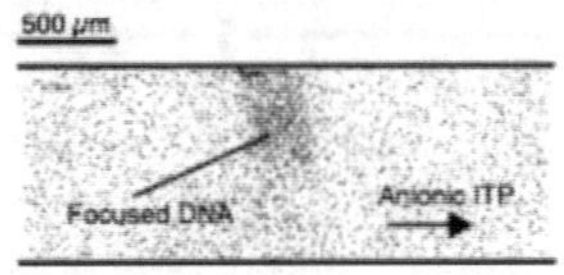

Figure 4. Fluorescent image of DNA labeled with TOTO-3 in the ITP zone. DNA migrates toward the anode.

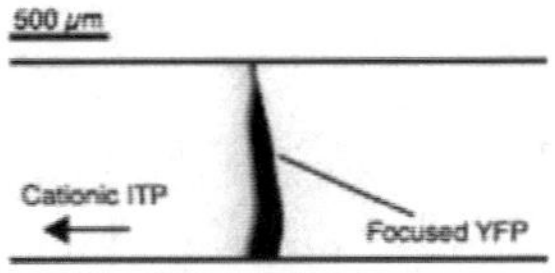

Figure 5. Fluorescent image of YFP in the cationic ITP zone. YFP migrates toward the cathode.

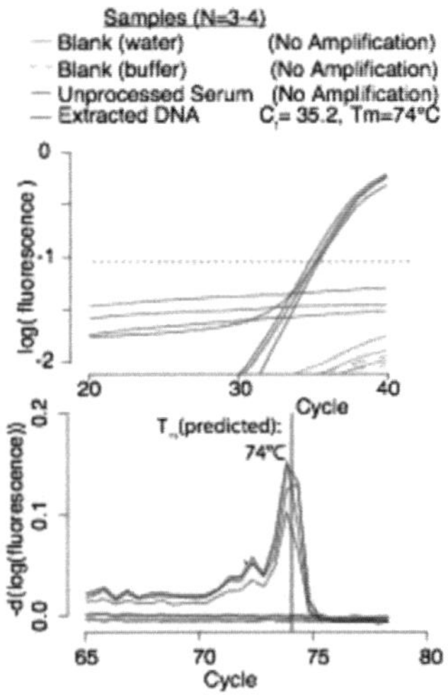

Figure 6. qPCR results: extracted nucleic acids were amplified using primers for a 201-bp section of the BRAC2 gene in the human genome. The extracted DNA samples amplified, while control samples of water and unprocessed serum did not. The resulting amplicons had a melting temperature of 74°C, which matches the prediction from the Promega amplicon melting tool.

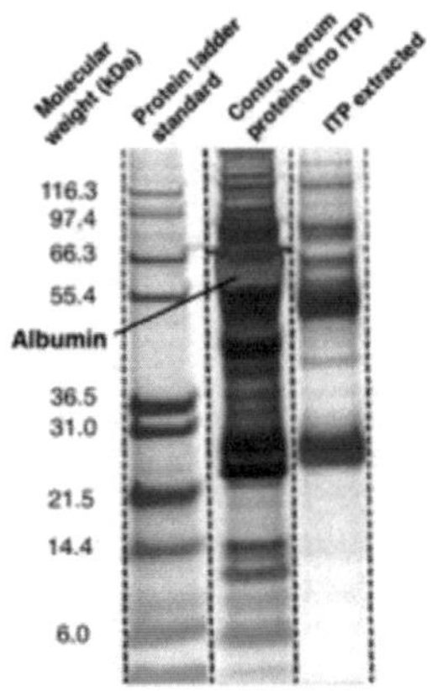

Figure 7. Silver-stained SDS-PAGE of original serum sample, and proteins extracted using ITP. We recover at least 17 visible protein bands over a range of molecular weights. Protein bands below 21.5 kDa are faint but detectable. Albumin is the most concentrated band in the serum sample, but is not visible in the selectively extracted sample.

REFERENCES

ADDIN EN.REFLIST 1. Persat, A., Marshall, L.A., Santiago, J.G., Purification of Nucleic Acids from Whole Blood Using Isotachophoresis. *Analytical Chemistry* **2009,** *81* (22), 9507-9511.

2. O'Driscoll, L., Extracellular Nucleic Acids and their Potential as Diagnostic, Prognostic and Predictive Biomarkers. *Anticanser Research* **2007,** (27), 1257-1266.

3. Butt, R.H., Pfeifer, T.A., Delaney, A., Grigliatt, T.A., Tetzlaff, W.G., Coorssen, J.R., Enabling Coupled Quantitative Genomics and Proteomics Analyses from Rat Spinal Cord Samples. *Molecular & Cellular Proteomics* **2007,** (6), 1574-1588.

4. Promega http://www.promega.com/resources/tools/biomath-calculators/

(320c) Direct Numerical Simulation of Electrokinetic Chaos Near Ion-Selective Surfaces

Davidson, S. M., Stanford University
Druzgalski, C. L., Stanford University
Andersen, M. B., Stanford University
Mani, A., Stanford University
Originally presented on: 11/5/2013 13:14:00 - 13:36:00

Many electrochemical cells involve aqueous electrolytes with charge transport through ion-selective surfaces, such as membranes or electrodes. Using direct numerical simulation (DNS) of the coupled Poisson-Nernst-Planck and Navier-Stokes equations in a multidimensional domain, we analyze the transport processes driving electrohydrodynamic chaos near these surfaces. Relevant model problems that we explore include an aqueous electrolyte between an ion-selective membrane and a stationary reservoir, and an aqueous electrolyte between electrodes. Electrode surfaces are modeled with realistic Butler-Volmer reaction kinetics and account for the Stern-layer capacitance. Our results indicate that as the voltage drop in the electrolyte exceeds ~10-100 thermal voltages, the system transitions to a regime involving fully chaotic vortices. In contrast to the previously known vortices due to electro-osmotic instability, the chaotic vortices are highly unsteady, with a multilayer structure involving multiple vortex sizes and broadband velocity spectra. The current-voltage predictions from our simulations successfully predict an over-limiting current regime consistent with recent experimental observations. The practical implications of these vortices include: enhanced mixing, enhanced transport beyond diffusion limitation, and reduced overpotential due to concentration gradients. We will discuss the relevance of turbulence theory tools for mathematical modeling of such regimes. Namely, we show that turbulent-diffusivity-type models are promising for prediction of the net transport rate and average concentration profiles in such systems.

Snita, D., Institute of Chemical Technology, Prague

Patel, N.,

Originally presented on: 11/5/2013 13:36:00 - 13:58:00

We developed a mathematical model describing traveling-wave electroosmotic micropumps to explain their rather poor ability to work against pressure loads. Unlike most of the authors, we use the combination of the Poisson-Nernst-Planck and the Navier-Stokes equations, without common simplifications (e.g., the slip boundary condition for velocity and the thin double layer approximation). The dynamical simulations allow a detail study of the energy transformations and the charging of the electric double layers. Using Matlab and COMSOL Multiphysics, we performed series of extensive parametric studies to determine the dependence of generated electroosmotic flow on the arrangement of the pump and on the characteristics of the driving ac-voltages. The results are consistent with our belief that all ac electroosmotic pumps tested so far were simply too large. According to our findings, the ac electroosmosis is likely to be suitable only at sub-micrometer scale, as the pump's ability to work against pressure loads diminishes rapidly when increasing the channel diameter.

(320e) Electrokinetic Translocation of Nanoparticles Through Nanopores Under Concentration Gradients

Hu, G.,
Wang, J.,
Originally presented on: 11/5/2013 13:58:00 - 14:20:00

We present experimental and theoretical studies of electrokinetic translocation of nanoparticles through solid-state nanopores with/without concentration gradients. Firstly, we used the Coulter counter principle (also known as resistive pulse method) to investigate the electrokinetic translocation of charged polystyrene (PS) nanoparticles (50- and 100-nm-diameter) through solid-state silicon nitride nanopores with diameters from 80 nm to 200 nm, respectively. The time-dependent electric current was recorded as the nanoparticles were driven across nanopores by externally applied voltages. The size of nanoparticles can be discriminated by the different magnitudes of current whereas the mean velocity of particle translocation can be estimated by measuring the duration of the current pulse. The translocation phenomena were further studied under different operating conditions, by changing salt concentrations, pH values, and salt types. In addition to electrophoresis and electroosmosis, a charged particle is also supposed to experience diffusiophoresis arising from the salt concentration gradients. Therefore, we conducted experiments to study how concentration gradients affect the particle translocation. The experiments were carried out at different salt concentration gradients, different sizes of the nanopores and the nanoparticles, and different types of the salts. In order to better understand the mechanisms of particle translocation through nanopores subject to concentration gradients, we further developed a continuum model that consists of the Poisson equation for the electric potential distribution, the Nernst-Planck equations for the ionic mass transport, and the Navier-Stokes equations for the fluid flow. We computed the position-dependent nanoparticle velocity as the nanoparticle moves through the nanopores. The resistive pulse of computational results is in good agreement with the corresponding experimental observations. Both the experimental measurements and computational results indicate that imposing a concentration gradient is an effective way to regulate translocation velocity of nanoparticles through solid-state nanopores. This technique can also be used to regulate the DNA translocation through nanopores for a better DNA detection.

(320f) Stern Layer Effect On the Field Effect Regulation of Surface Charge Property and Electroosmotic Flow in a Silica Nanochannel

Yeh, L. H.,

Originally presented on: 11/5/2013 14:20:00 - 14:42:00

Electrokinetic transport of ions, fluid, and (bio)particles in micro- and nano-fluidics has attracted considerable attention because of its widespread applications ranging from biosensing to the separation of (bio)molecules. Many of these applications require active control of the surface charge property at the solid/liquid interface of the micro/nanochannel. To achieve this end, field effect transistors (FETs), consisting of gate electrodes embedded beneath dielectric micro/nanochannel walls, have been developed to actively tune the surface charge property. In this study, active control of the surface charge property and electroosmotic flow (EOF) in a silica-based nanochannel by a field effect transistor (FET) is analyzed for the first time with the consideration of the Stern layer effect. Approximate expressions for estimation of the surface charge property and EOF velocity have been derived and validated by comparing to experimental data available from the literature. Results obtained show that field effect regulation of the surface charge property and EOF velocity depend highly on the surface capacitance of the Stern layer and background solution properties including pH and salt concentration. The Stern layer effect becomes appreciable when the magnitude of the surface potential is high, occurring at high pH and/or gate potential imposed.

2013 Annual Meeting of the American Electrophoresis Society (AES)

(320g) An Electrokinetic Probe of DNA Binding Interactions Via Resonant Entropic Trapping

Shi, N., Texas A&M University

Ugaz, V. M., Texas A&M University

Originally presented on: 11/5/2013 14:42:00 - 15:04:00

A host of biological processes are orchestrated by complexation between DNA and binding agents (small molecules, proteins, adducts), and a fundamental understanding of these interactions is critical to unlocking their envisioned functionality. Nanopore-based methods have recently demonstrated potential to screen binding events and infer their kinetics, but conformational changes induced in the DNA complex upon binding must be quantitatively mapped in order to ascertain their biological function. These structural details can currently be elucidated by single molecule force measurements employing optical or magnetic tweezers and atomic force microscopy (AFM), but the corresponding experimental platforms are challenging to parallelize and analysis is inherently restricted to large sized biomolecules that can be readily imaged (i.e., λ-phage DNA or longer, contour lengths ~ μm).

Here we show how these limitations can be uniquely overcome by exploiting resonant entropic trapping (RET) imposed when DNA molecules experience confinement within a nanoscale pore network of comparable size to their equilibrium random coil conformation. This approach enables fundamental structural parameters associated with binding and complexation (contour and persistence lengths), including adduct formation, to be finely resolved in DNA much smaller than the size limits of conventional measurement techniques. In this way, we are able to gather structural information associated with DNA-drug complexation by performing a single microchip electrophoresis experiment (~20 min run time). RET can be accessed within the inherently nanoporous architecture of conventional polymeric hydrogels, eliminating the need to construct artificial nanofabricated topologies and making it possible for untethered DNA of arbitrary length to be rapidly analyzed in a convenient highly scalable format.

(358a) Directed Dielectrophoretic Assembly of Thin Highly Organized Photoreactive Biocoatings of Cyanobacteria

Bernal, O. I., North Carolina State University
Flickinger, M. C., Golden LEAF Biomanufacturing Training and Education Center (BTEC)
Velev, O. D., North Carolina State University
Originally presented on: 11/5/2013 15:15:00 - 15:33:00

The use of dielectrophoresis (DEP) for the fabrication of monolayer coatings of photoreactive cyanobacteria and adhesive polymers on flexible polyester sheets is being investigated to optimize photoreactivity, gas transport, light scattering and self-shading in future thin coating photobioreactors by precisely controlling cell packing and orientation. DEP is being used to fabricate 1-cell thick coatings of *Synechococcus PCC 7002* to study light absorption and scattering, cell packing, and reactivity after drying and rehydration. This organism can be genetically modified to produce fuel precursors from CO_2 with simultaneous energy capture from sunlight and O_2 production. Adhesion is triggered by priming a surface-activated polyester substrate with a layer-by-layer (l-b-l) assembled film of poly(allylamine hydrochloride) (PAH) and poly(styrene sulfonic acid) (PSS) [general composition $(PAH\text{-}PSS)_3PAH$]. The effects of voltage, frequency, electrolyte concentration and pH during the DEP assembly demonstrate the means to fabricate viable chains and monolayers of *Synechococcus* in suspension, and deposit on polyelectrolyte-coated polyester sheets permanent ordered biocoatings that remain aligned after the electric field is turned off. Photoreactivity (measured as reflectance, absorptance and transmittance using an integrating sphere) of DEP coatings of *Synechococcus PCC 7002* show significantly decreased light scattering and enhanced transmittance indicating reduced self-shading compared to suspension cultures. Analysis of confocal laser scanning microscopy (CLSM) images allows accurate measurement of cell dimensions for predicting optimal cell packing density and for electrostatic simulations of chain assembly. Numerical simulations of chain and monolayer assembly in suspension show that field intensity is strongest between adjacent cells in chains and drives chains together to form monolayers. Total electrostatic energy calculations predict energetically favorable intermediate cell ordering motifs and reveal mechanisms of coating assembly by comparing simulation results with experimental observations. Future work will be focused on microstructure characterization of the DEP coatings, development of methods for photoreactivity analysis and development of layer-by-layer (l-b-l) DEP-based methods for fabrication of multi-layered coatings combining different types of photosynthetic organisms. These methods will be used to construct highly reactive, robust, structured cellular biomimetic materials capable of being dried and when rehydrated harvest light in a broad range of wavelengths and remain reactive for 1000s of hours.

(358b) Co-Electrophoretic Deposition of Composites: Understanding Deposition Mechanisms

Pascall, A. J., Lawrence Livermore National Laboratory
Mihaljevich, B., Lawrence Livermore National Laboratory
Sullivan, K. T., Lawrence Livermore National Laboratory
Kuntz, J. D., Lawrence Livermore National Laboratory
Originally presented on: 11/5/2013 15:33:00 - 15:51:00

Electrophoretic deposition (EPD) is a process in which a suspension of colloidal particles is driven to deposit onto the surface of an electrode under an applied electric field. EPD has been used industrially for decades as a coating process, for instance, to deposit ceramic on metal for thermal and diffusion barriers. Recently, EPD has been employed to fabricate functional and functionally graded materials. These applications entail depositing a composite of at least two different materials. A method for accomplishing this is to co-suspend the materials of interest in a solvent and deposit both simultaneously. However, it has been observed that, in general, the composition of the coating does not match the composition of the suspension. Two mechanisms have been proposed account for this: each material has a different electrophoretic mobility and hence deposits at a different rate, or the materials form agglomerates in the bulk, which then deposit by EPD. Here, we present a study in which we systematically vary the composition of the suspension and deposition parameters to determine the effects on final deposit composition in an attempt to differentiate between the two mechanisms.

(358c) Assembly of Janus Particles In Combined Electric and Magnetic Fields

Kretzschmar, I., The City College of New York, The City University of New York
Originally presented on: 11/5/2013 15:51:00 - 16:09:00

Gold-capped Janus particles have been shown to assemble into staggered chains upon exposure to an AC electric field. The AC electric field causes the polarization of the metal caps resulting in a dipole that is positioned at the base of the cap and is aligned with the external electric field. The particles assume an orientation within the staggered chain in which the caps of two adjacent particles touch. Simulations show that this configuration allows for the energetically most favorable dipole-dipole interaction. Recently, we have prepared Janus particles with varying iron oxide caps. One of the iron oxides, Fe3O4, exhibits both conducting and magnetic properties. Exposure of the Fe3O4 Janus particles to a magnetic field leads to close-packed double chains, while staggered chains are formed in an AC electric field. This finding opens up the possibility to study chain dynamics, if conditions can be found in which the relative stability of the two minima can be varied through the external field strength. In this talk, we will report on the behavior of Fe3O4 capped Janus particles in the presence of combined electric and magnetic fields and present a simple model that explains it.

2013 Annual Meeting of the American Electrophoresis Society (AES)

(358d) The Impact of Geometric Anisotropy On Colloids Under Electric Fields

Ma, F., Colorado School of Mines

Wang, S., Colorado School of Mines

Wu, D. T., Colorado School Of Mines

Wu, N., Colorado School of Mines

Originally presented on: 11/5/2013 16:09:00 - 16:27:00

Colloids possessing anisotropic interactions could potentially assemble into a much wider range of crystalline arrays and meso-structures than spherical particles with isotropic interactions. Here, we specifically investigate the impact of geometric anisotropy on the assembly of colloidal dimers on conducting substrates under applied electric fields. By systematically tuning the size ratio between two lobes on dimers, we have found that interactions between dimers are strongly dependent on the their relative orientations. For example, the interaction between lying and standing dimers on the substrate is attractive. When all dimers stand on the substrate, the interaction between neighboring dimers with alternating orientations is also attractive. Otherwise, it is repulsive. Such kind of orientation-dependent interactions generate a good variety of new structures, such as chiral clusters and dimer crystals with alternating orientations. We will discuss the physical origin of those orientation-dependent interactions and the impacts of experimental conditions such as the ionic concentrations and surface charges. Our numerical model based on electrostatics agrees well with experimental observations and provide further insights on electric-field assisted assembly of anisotropic particles.

(358e) Effects of Microtopography On Two-Dimensional Electrokinetic Patterning of Colloids On An Electrode Surface

Williams, S. J., University of Louisville
Originally presented on: 11/5/2013 16:27:00 - 16:45:00

Colloid self-assembly techniques are beneficial due to their ability to create organized architectures from a disordered suspension. Colloids have been used to enhance material properties including mechanical, heat transfer, or optical characteristics. There are inherent disadvantages of established colloid assembly techniques (solvent evaporation, sedimentation, and layer by layer deposition), namely the rate of assembly and the precision/resolution of the desired pattern. AC electrokinetic techniques, including rapid electrokinetic patterning (REP), have demonstrated rapid colloidal assembly and patterning.

This work explores the use of templates on an electrode surface to guide and retain colloids patterned on the electrode's surface. A parallel-plate electrode is used using two indium tin oxide (ITO) substrates (initial ITO thickness is approximately 500 nm). Simple micrometer geometries are etched into one ITO electrode. 1.0 micrometer particles are used to demonstrate the effect of microtopography on 2D colloidal patterning. A combination and balance of uniform and non-uniform electrokinetics, particle-particle interactions, and particle-electrode interactions result in improved colloidal pattern resolution.

2013 Annual Meeting of the American Electrophoresis Society (AES)

(358f) Electrolyte Dependence of Particle Motion Near An Electrode During AC Polarization

Wirth, C. L., Katholieke Universiteit Leuven

Sides, P. J., Carnegie Mellon University

Prieve, D. C., Carnegie Mellon University

Originally presented on: 11/5/2013 16:45:00 - 17:03:00

The directed motion of colloidal particles on a planar electrode has been studied for more than two decades because of its importance to the fabrication of colloidal crystals, the separation of colloidal dispersions, and the high throughput evaluation of electrocatalysts. The phase angle between the imposed ac electric field and the oscillations in particle height is the key parameter governing the sign of inter-particle force and, consequently, the equilibrium microstructure of the colloidal crystal. The phase angle depends on a number of experimental parameters, including the strength and frequency of the imposed electric field. Surprisingly, the phase angle also depends on the dispersing electrolyte. Negatively charged polystyrene spheres 5.7 μm in diameter dispersed in 0.15 mM KCl or $NaHCO_3$ have a phase angle > 90°, but particles dispersed in 0.15 mM NH_4OH or KOH have a phase angle < 90° despite all other experimental parameters being identical to the previous case. Electrolyte dependence arises in both the polarization of the particle's diffuse layer and the conductivity of the dispersing electrolyte. The first *a priori* model for electrolyte dependent particle motion normal to an electrode during ac polarization will be presented.

curing process, dispersion time and rate ,were experimentally investigated. And the result showed that the optimum conditions are as followsÂ □ Fthe ratio of gelatin and gum arabic to electrophoretic fluid was 1:4(w:w), the microcapsule wall should be cured for 70 min at pH 8 to 11 (T=45°C) ,the time and rate to disperse the electrophoretic fluid was 20 min and 800 rpm,respectively.The particle size of the electrophoretic microcapsule prepared in such conditions was well-distributed in 30Â □`60μm. it is found that the particles represent electrophoretic behavior when the field strength was $4000V\cdot m^{-1}$. The electrophoretic display devices composed of these prepared microcapsules have good display properties when driven by the active or passive matrix driving.

(358g) Predicting the Disorder-Order Transition of Dielectrophoretic Colloidal Assembly With Dielectric Spectroscopy

Beltramo, P. J., University of Pennsylvania
Furst, E. M., University of Delaware
Originally presented on: 11/5/2013 17:03:00 - 17:21:00

The self-assembly of colloidal building blocks into ordered, periodic structures using directing electric fields is a rapid, highly scalable, and potentially low-cost route for manufacturing functional nanomaterials and devices. The dielectrophoretic assembly of colloidal suspensions into crystalline arrays follows a master scaling that collapses the disorder-order transition as a function of field strength, frequency and particle size. This master scaling has been verified for particle diameters ranging from $2a = 200$ nm to 3 mm by light scattering (Lumsdon et al., *Langmuir* 20, 2108-2116, 2004; McMullan and Wagner, *Langmuir* 28, 4123-4130, 2012), laser tweezer pair interaction measurements (Mittal et al., *J. Chem. Phys.* 129, 064513, 2008) and small-angle neutron scattering (McMullan and Wagner, *Soft Matter* 6, 5443-5450, 2010). In this work, we reconcile the established empirical phase diagram with direct measurements of the colloid polarizability using dielectric spectroscopy. By varying volume fraction, particle size and ionic strength, we confirm the disorder-to-crystal electric field assembly phase diagram. This represents an alternative means to search for optimal self-assembly conditions that is not limited by particle size, shape, chemistry or solvent characteristics.

(358h) The Preparation of Environmental Friendly Gelatin-Gum Arabic Microcapsule for Electrophoretic Display

Liu, H., Tianjin University
Wang, S., School of Chemical Engineering and Technology, Tianjin University
Li, X., School of Chemical Engineering and Technology, Tianjin University
Xiao, Y., School of Chemical Engineering and Technology, Tianjin University
Originally presented on: 11/5/2013 17:21:00 - 17:39:00

The Preparation of Environmental Friendly Gelatin-Gum Arabic Microcapsule for Electrophoretic Display

Hongli Liu ,Shirong Wang,Xianggao Li ,Yin Xiao*

(School of Chemical Engineering and Technology, Tianjin University, Tianjin,300072, China)

Abstract Electrophoretic display (EPD),which present images as the result of the particles with opposite charges moving in the electric field ,is know as the technology to relize the reflective display.And it has attracted more and more attention due to its low cost, low weight, good flexibility, wide viewing angles and low power consumption.

In order to get high response rate and contrast ratio electrophoretic display,the black/white particles together with charge control agent and dispersant et al.were coated in the gelatin-gum arabic microcapsule. Microencapsulated electrophoretic ink possesses the merits of flexibility,bistablity, extremely low power consumption, high contrast and reflectivity, as well as wide view angle. So it was successfully used in the preparation of electronic paper.

the choice of dispersion medium is especially critical for the electrophoretic fluid . At present, tetrachlorethylene is commonly used as dispersion medium, while it is not only harmful to human beings because of the effect of stimulation and anesthesia but also contaminates the environment by polluting underground water. In this article, isopar L ,which is isoparaffin,was selected as the dispersion medium instead of tetrachlorethylene. It is not only nearly non-toxic to human beings but also environmental friendly because of its low aromatic concentration and volatility.

The electrophoretic microcapsules containing carbon black and TiO_2 particles dispersed in isopar L were prepared by complex coacervation between gelatin and acacia. Carbon black and TiO_2 particles in the electrophoretic microcapsules were prepared by ball milling. Various factors affecting the properties of the prepared microcapsules, such as the ratio of gelatin and gum arabic to electrophoretic fluid , pH value and curing process, dispersion time and rate ,were experimentally investigated. And the result showed that the optimum conditions are as followsÂ □ Fthe ratio of gelatin and gum arabic to electrophoretic fluid was 1:4(w:w), the microcapsule wall should be cured for 70 min at pH 8 to 11 (T=45°C) ,the time and rate to disperse the electrophoretic fluid was 20 min and 800 rpm,respectively.The particle size of the electrophoretic microcapsule prepared in such conditions was well-distributed in 30Â □`60μm. it is found that the particles represent electrophoretic behavior when the field strength was $4000 V\cdot m^{-1}$. The electrophoretic display devices composed of these prepared microcapsules have good display properties when driven by the active or passive matrix driving.

(392i) Electrokinetic Behavior of Large Polystyrene Particles in Insulator Based Dielectrophoresis

Gencoglu, A., Rochester Institute of Technology
Gulati, N., Rochester Institute of Technology
Bruening, D., Rochester Institute of Technology
La Londe, A., Rochester Institute of Technology
Koppula, K. S., Rochester Institute of Technology
Lapizco-Encinas, B., Rochester Institute of Technology

Dielectrophoresis is a microscale electrokinetic technique that has been successfully employed for the manipulation and assessment of a wide array of bioparticles, from proteins, to DNA to cells. This work is focused on insulator-based dielectrophoresis (iDEP), a technique where dielectrophoretic forces are generated by creating non-uniform electrify fields using insulating structures embedded between two external electrodes. It was recently observed that "larger" (>5 μm diameter) polystyrene spheres did not exhibit a clear negative dielectrophoresis (DEP) behavior when working with a microfluidic channel with insulating posts in insulator-based dielectrophoresis (iDEP) experiments. DEP theory predicts that larger particles will experience much stronger trapping than smaller particles, which have been easily manipulated under identical conditions. Cells in the 5-10 μm range have also been successfully manipulated using iDEP. Understanding the deviation of large polystyrene particles from predictions of current DEP theory may lead to novel applications and methods for DEP based manipulation and separation of particles.

(392j) Size Based Separation of Lipid Droplets Using Insulator-Based Dielectrophoresis

Gencoglu, A., Rochester Institute of Technology
Meyers, A.,
Dalhaimer, P., University of Tennessee
Lapizco-Encinas, B., Rochester Institute of Technology

Cells store neutral lipids in cytosolic lipid droplets, which are hypothesized to form from the endoplasmic reticulum. Droplets are often "called obesity" organelles because they occupy the vast majority of the volume of adipocytes in white adipose tissue. Much information on the roles of lipid droplets has been obtained by proteomic and lipidomic analysis of isolated droplets. However, lipid droplets harvested from cultured cells can only be analyzed in aggregate. Here, we present size based sorting of lipid droplets using insulator-based dielectrophoresis (iDEP). Submicron lipid droplets from yeast cells and lipid cells from human hepatocytes and human adipocytes in the 1-10 μm range were sorted according to their sizes in a microchannel with insulating posts using insulator-based dielectrophoresis (iDEP). DEP refers to the movement of particles, due to polarization effects when particles are exposed to non-uniform electric fields. In particular, in iDEP, these electric field gradients are generated by employing insulating structured between two external electrodes. By carefully selecting operating conditions and microdevice design, it was possible to concentrate and sort a sample of lipid droplets.

2013 Annual Meeting of the American Electrophoresis Society (AES)

(392d) Assembly of "Anisotropic" Colloidal Dimers and Spheres Under Applied Electric Fields

Ma, F., Colorado School of Mines

Wang, S., Colorado School of Mines

Wu, D. T., Colorado School Of Mines

Wu, N., Colorado School of Mines

Colloids possessing anisotropic interactions could potentially assemble into a much wider range of crystalline arrays and meso-structures than spherical particles with isotropic interactions. Here, we investigate the impact of geometric anisotropy on the assembly of colloidal dimers on conducting substrates under external electric fields. By systematically tuning the size ratio between two lobes on dimers, we have found that interactions between dimers strongly depend on their relative orientations. For example, the interaction between lying and standing dimers on the substrate is attractive. When all dimers stand on the substrate, the interaction between neighboring dimers with alternating orientations is also attractive. Otherwise, it is repulsive. Such kind of orientation-dependent interactions generate a good variety of new structures, such as chiral clusters and dimer crystals with alternating orientations. We will discuss the physical origin of those orientation-dependent interactions and the impacts of experimental conditions such as the ionic concentrations and surface charges. Our numerical model based on electrostatics agrees well with experimental observations and provide further insights on electric-field assisted assembly of anisotropic particles.

In addition, the assembly of isotropic spherical colloids under applied electric fields is investigated. By applying an external AC electric field, we show that the apparently isotropic particles experience "anisotropic" interactions. New types of sphere-packing have been observed within a previously unexplored experimental regime: low salt concentrations ($<10^{-3}$M) and low frequency regime (100 Hz to 10 kHz). At low particles concentrations, a family of well-defined oligomers, ranging from 3 to 10 was observed. At high particles concentrations, the colloidal clusters will further assemble and connect themselves into a good variety of two-dimensional non-close-packed networks. We attribute these new types of non-planar structures to the competition among double layer repulsion, dielectrophoretic attraction, and dipolar interactions. The double layer and in-plane dipolar repulsion could make bottom particles in the clusters separate from each other. While the out-of-plane dipolar attraction and particle-substrate dielectrophoretic attraction could be responsible for the formation of the clusters, i.e., the top central sphere is associated with the bottom spheres. The effects of salt concentration and frequency on the geometry of those colloidal molecules will be discussed. These non-close-packed structures could be used as building blocks for making photonic crystals and plasmonic structures.

(392k) Using Low Frequency Electrical Signals for Particle Separations with Dielectrophoresis

Gencoglu, A., Rochester Institute of Technology
Olney, D., Rochester Institute of Technology
La Londe, A., Rochester Institute of Technology
Koppula, K. S.,
Lapizco-Encinas, B., Rochester Institute of Technology

In insulator-based dielectrophoresis (iDEP), particles are trapped and immobilized when the dielectrophoretic force acting on them dominates over other transport mechanisms, such as by electroosmotic flow (EOF). Thus, particles are released from the “dielectrophoretic” raps when EOF dominates over the DEP. Cyclical electrical signals were used to reduce and fine tune EOF transport by allowing iDEP separations using lower applied electric potentials than reported before. The cyclical signals reduced net EOF since EOF direction depends on the electric field direction. DEP direction does not change with the direction of the electric field, and was not weakened by a cyclical signal. Polystyrene spheres of different sizes were separated in a microchannel with cylindrical insulating structures 450 μm diameter and spaced 500 sym center to center. Initially, a sinusoidal signal between +400 and -400 V was applied, and minimum value was gradually increased, while the maximum value was kept constant at +400 V. This method achieves iDEP separations at lower applied voltages than those reported in previous work for similar particle mixtures.

(392e) X-Ray and Raman Transparent Solvent Resistant Microfluidic Platforms to Screen Solid Forms of Pharmaceuticals

Goyal, S., University of Illinois at Urbana Champaign
Economou, A., University of Illinois at Urbana Champaign
Lee, G., University of Illinois at Urbana Champaign
Gong, Y., AbbVie Inc.
Zhang, G. G. Z., AbbVie Inc.
Kenis, P. J. A., University of Illinois at Urbana Champaign

We present microfluidic platforms for screening solid form (salts, co-crystals, and polymorphs) of pharmaceutical parent compounds (PC). Success in finding a crystalline solid form of a PC with optimized CMC properties using conventional screening procedures is limited due to the limitations of the number of experimental conditions that can be investigated while utilizing small quantities of PCs available ($\approx$ 10 mg) in the early stages of drug development. Microfluidic technology allows screening with reduced sample volumes of PC and precipitants (salt or cocrystal formers or antisolvents) solutions [1] and thereby enable screening of multiple conditions using the limited amount of PC. This will enable early identification of all possible solid forms of PCs and help reduce the time and money invested in the solid form development.

To date, we have reported on hybrid polymer-based microfluidic platforms that permit combinatorial mixing of PC and precipitants solutions on-chip in arrays of 24/48, 50 to 200 nL wells, which is a drastic reduction in the volume of PC needed per condition compared to traditional approaches (~5 to 100 ul per condition [2]) [3-6]. These platforms were compatible with mild organic solvents and water, and enabled identification of the solid form crystallized on-chip via on-chip Raman spectroscopy. We screened for solid forms of PCs employing free interface diffusion (FID) mixing [3,5], antisolvent addition (AS) [4], and solvent evaporation [6] modes of crystallization employing the polymer-based platforms.

Here we present the design, fabrication, and application of solvent resistant microfluidic platforms for solid form screening of PCs. These newly developed hybrid polymer-based microfluidic platforms are compatible with a much wider range of solvents including chloroform, toluene, and hexanes, thereby enabling investigation of a much wider range of conditions on-chip as well as allow a better control on the supersaturation levels attained on-chip. We employed these platforms to screen for solid forms of model compounds including theophylline, piroxicam, and carbamazepine. On-chip Raman and X-ray analysis was employed to identify and distinguish different solid forms crystallized on-chip. These platforms have the potential to expedite the drug development process by enabling solid form screening at the early stages of drug development.

References

[1] C. Hansen, E. Skordalakes, J. Berger, S. Quake, Proc. Nat'l Acad. Sci., 2002, 99, 16531-6.

[2] S. Talreja, P.J.A. Kenis, et al. J. Appl. Crystallography, 2005, 38, 988-995.

[3] M.R. Thorson, S. Goyal, P.J.A. Kenis, et al. Lab Chip, 2011, *11, 3829-3837*

[4] M.R. Thorson, S. Goyal, P.J.A. Kenis, et al. CrystEngComm, 2012, 14, 2404-2412

[5] S. Goyal, P.J.A. Kenis, et al. Crystal Growth & Design, 2012, 12 (12), 6023-6034

[6] S. Goyal, P.J.A. Kenis, et al. Lab Chip, 2013, 13, 1708-1723

2013 Annual Meeting of the American Electrophoresis Society (AES)

(392g) Cell Lysis in Microfluidic Devices Employing DC Electric Currents

Gallo-Villanueva, R. C., Tecnológico de Monterrey

Rodriguez-Lopez, C. E., Tecnologico de Monterrey

Díaz-de-la-Garza, R. I., Tecnologico de Monterrey

Lapizco-Encinas, B., Rochester Institute of Technology

Martinez-Chapa, S. O., Tecnológico de Monterrey

Electrokinetics is a widely employed technique for particle manipulation in microdevices. Combination of electrophoretic, electroosmotic, and dielectrophoretic, among other forces are used with AC and/or DC electric fields inside microchannels for different applications. In this work, cell integrity of *E. coli* was evaluated in a microchannel after applying DC electric fields, as a method for cell lysis for DNA recovery. The use of surfactants along with electric field gradients, created by the inclusion of insulating posts in the channel, caused cells to lose viability and release cytoplasmic material to the medium. This method can be used as an alternative for cell lysis and applied to protein, RNA, or DNA recovery.

(392h) Fabrication of 3D Electrodes for Electrorotation Experiments

Guiducci, C., EPFL

Fabrication of 3D Electrodes for Electrorotation Experiments

Samuel Kilchenmann* and Carlotta Guiducci*

*École Polytechnique Fédérale de Lausanne, SWITZERLAND

Introduction

Electrorotation (ROT) is a powerful technique that has been used to characterize and differentiate different particles through the extraction of their electrical parameters. Through the study of a particle's response to an electric field at specific frequencies, it is possible to extract information such as the membrane integrity of biological particles. In order to achieve high accuracy of the extracted parameters, a rotating electric field with high linearity needs to be generated. For this purpose, Goater et al. introduced bone-shaped electrodes [1]. Most devices used for electrorotation consist of planar electrodes that result in electrical fields in z-direction, which causes out of focus movements and makes trapping and observation of the particles difficult. Different groups tried to overcome this problem by designing chips with octopole cages that trap and rotate cells at the same time [2] or by combining the ROT devices with laser tweezers for trapping [3]. We introduce a novel fabrication process for three-dimensional microelectrodes that can be used for electrorotation experiments. The advantages of our design are that the generated electric field does not have any z-component and thus no out-of-focal-plane movements are exerted on the particle. Moreover, the electrodes serve as a kind of mechanical trap for the particles. Further advantages of the presented fabrication process are that the microfluidic structures can be aligned with photolithographic precision and that it is possible to work with bulky and short-working distance lenses.

Simulation

The electrical fields that are generated by 3D and planar electrodes were simulated using Comsol Multiphysics. In both cases a bone shaped design of the electrodes was used for simulation. From the top views in Fig. 1 it can be seen that both designs lead to highly linear fields in the x-y plane, however, in the case of the planar electrodes the generated field lines have components in z-direction as well. A further advantage of 3D electrodes is that local heating effects can be reduced by about an order of magnitude, while keeping in the center the same field strengths as with the planar electrodes. This is due to the higher electrode surface that is used to apply the potential, which leads to a reduction of the current density.

Microfabrication

The microfabrication process is shown in Figure 2. We start with a silicon wafer passivated by Si_3N_4 and SiO_2 layers with thicknesses of 200 nm each. First, a photoresist is applied by spin coating and is structured by photolithography. The oxide and the nitride layers are etched and the design is transferred 300 μm deep into the silicon by a Bosch process. Then, a 10 μm thick SU-8 layer is patterned by photolithography. For better adhesion of the metal on the SU-8, the resist is hardbaked at 135°C and plasma activation is done before sputtering a Ti/Pt layer onto the structures. Afterwards, a dry film resist is laminated on the wafer and patterned by photolithography. The resist serves as etch mask for plasma etching of the metal layer. After stripping of the photoresist, a second SU-8 layer to pattern the microfluidic channels is applied. To planarize the SU-8 layer that is spun onto the high topographic surface, chemical mechanical polishing (CMP) is performed after the post exposure bake and before the development of the resist. This strategy is chosen that the undeveloped resist protects the access pads during CMP. Afterwards, the SU-8 is developed and the microfluidic accesses are opened up by etching the backside oxide and nitride layers and by grinding the wafer down to 300 μm from the backside. Finally, the wafer is diced and a coverslip is bonded onto the chips, where the SU-8 serves as adhesive. In Figure 3 we show a picture of such an electrorotation chip (a) and SEM images of (b) the 3D electrodes without glass coverage of the microfluidic channels and (c) the cross section of the whole chip with glass coverage.

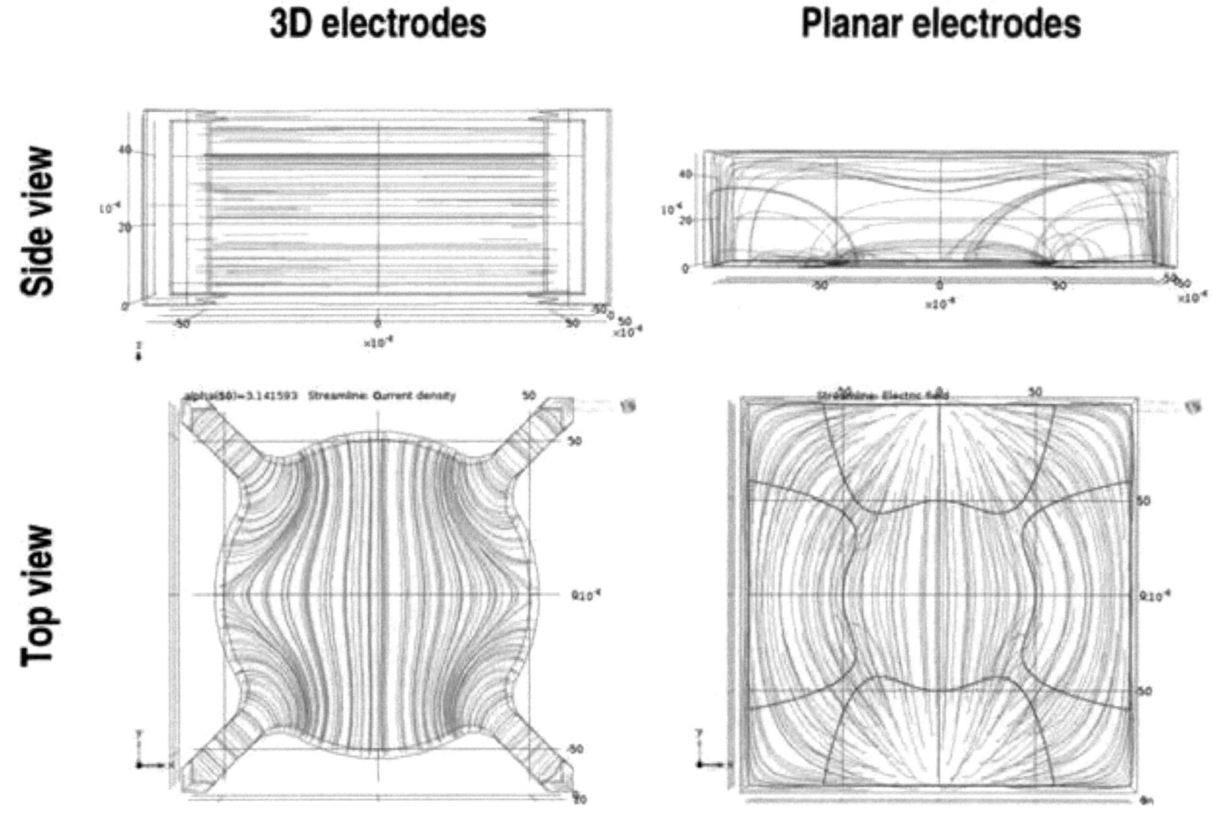

Figure 1 Comsol simulations of the electric field for planar and 3D electrodes

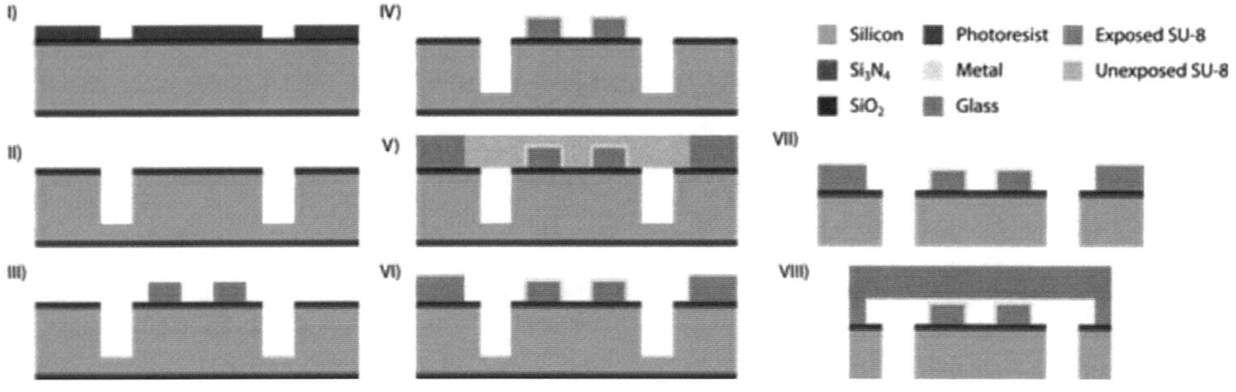

Figure 2 Fabrication process of the electrorotation chips with 3D electrodes

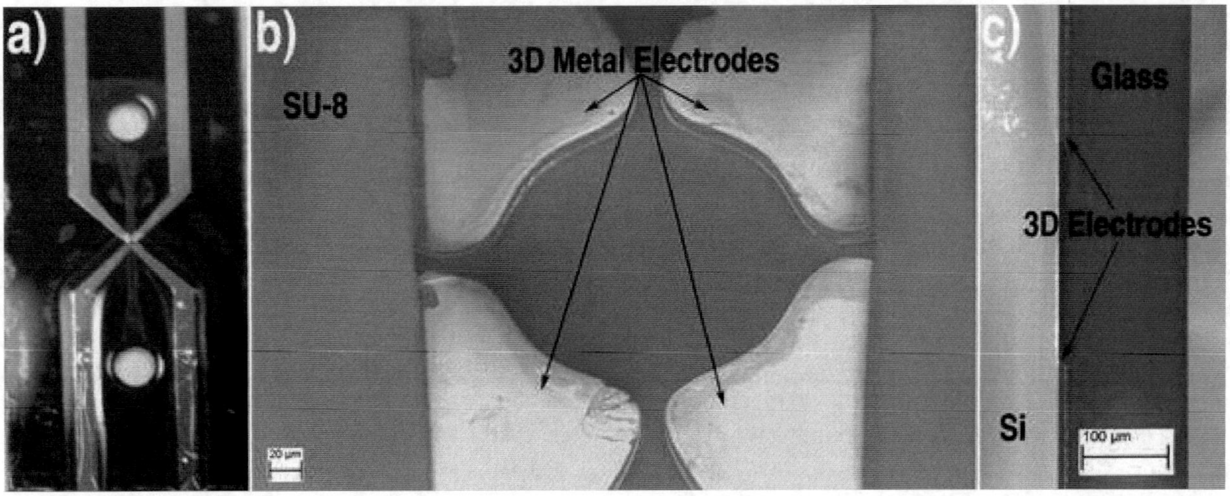

Figure 3: (a) Electrorotation chip ; (b)SEM of the 3D metal electrodes; (c)SEM of the cross-section of a chip with the glass coverage.

1. Dalton, C., et al., *Parasite viability by electrorotation.* Colloids and Surfaces a-Physicochemical and Engineering Aspects, 2001. 195(1-3): p. 263-268.

2. Reichle, C., et al., *Electro-rotation in octopole micro cages.* Journal of Physics D-Applied Physics, 1999. 32(16): p. 2128-2135.

3. Schnelle, T., et al., *Combined dielectrophoretic field cages and laser tweezers for electrorotation.* Applied Physics B-Lasers and Optics, 2000. 70(2): p. 267-274.

(392a) Bioparticle Separation in An Insulator-Based Dielectrophoretic Microchannel

Jones, P. V., Arizona State University
Hayes, M. A., Arizona State University

A significant and wide-ranging body of work is focused on rapid bioparticle characterizations. Clinical diagnosis will be directly impacted by innovations in this field because it will enable rapid, on-site bioanalysis, improving the availability, accuracy, and scope for these tests. For bioparticle analysis, microfluidic electrokinetic approaches provide unique advantages, including short analysis times, microliter sample and reagent volumes, potentially low cost, and portability. The work presented here explores a separations-based strategy for isolating and concentrating intact microorganisms within a single, continuous microchannel.

Using DC insulator-based dielectrophoresis in a converging, sawtooth-patterned microchannel creates multiple and distinct bioparticles traps. This channel design enables localized isolation and concentration of a wide variety of microbes and other particles based characteristic differences in their physical properties. Various targets have been captured and concentrated within the device, including human blood cells and mature amyloid protein fibrils. Recent work has focused on using this approach to differentiate very similar targets, such as different strains of a single bacterial species. This approach has demonstrated the ability to distinguish three serotypes of live *Escherichia coli*, indicating the potential this technique holds in terms of both separatory power and diagnostic applications.

(392f) Brownian Dynamics Simulations of Electrophoretic DNA Separations in a Conducting Post Array

Chen, C. A., National Taiwan University
Hsieh, C. C., National Taiwan University

We use Brownian dynamics simulations and computational fluid dynamics to exam the electrophoretic DNA separations through a conducting post array. Electrophoretic separation of DNA through a post array has been heavily investigated in both experiments and simulations. However, the posts were assumed to be either insulated or having the same conductivity as the buffer solution. In this study, we exam the idea that conducting posts may result in better DNA separation. Since the electric field lines concentrate around the more conductive posts, DNA molecules were expected to have higher probability to collide with the posts and to experience hooking and unhooking events more frequently, leading to better separation. In our simulations, we set the posts to have different degrees of conductivity higher than that of the buffer solution and analyze the average electrophoretic mobility and dispersion coefficient of lambda-DNA (48.5 kbp) and T4GT7-DNA (166 kbp) as they move through the array. However, the simulation results show that DNA separation has not been improved in spite of the higher collision rate. Moreover, in contrast to the observation in the insulated post array, T4GT7 DNA was found to move faster than lambda-DNA. We will explain these rather unexpected results with more detailed analysis.

2013 Annual Meeting of the American Electrophoresis Society (AES)

(392c) An Ipad-Based Brownian Dynamics Simulator for Electrokinetics in the Classroom

Shi, N., Texas A&M University

Ugaz, V. M., Texas A&M University

Tablet-based computing platforms offer enormous potential to provide immersive education experiences by virtue of their unique combination of a graphical environment and touch screen interface. Here we describe our efforts to exploit these features by porting a conventional Brownian dynamics simulation into an Apple iPad application that provides a visual representation of polymer coil size, relaxation phenomena, Brownian motion, and transport under an external driving force (i.e., electrophoresis).

Our app employs a coarse-grained bead-spring model of the polymer chain, enabling the user to customize the number of beads and their positions to define an initial polymer conformation. The user can choose between infinite and finite extension models for the connecting springs. The app's core simulation module provides two functions. First, in the free solution simulation mode, it mimics the polymer's Brownian evolution toward an equilibrium state. Key parameters continuously reported during the simulation include radius of gyration Rg, end-to-end length L, largest and smallest spring lengths, and contour length (determined from the average spring length). Secondly, the app is capable of simulating the polymer's response when a user moves individual beads by touching the screen, with force proportional to the speed of the user's finger swipe on the screen. This capability can be used to visualize the process of polymer stretching (single bead drag) or electrophoretic transport (whole polymer drag).

The app-based format makes it possible to visualize polymer dynamics in an appealing way and offers a useful tool to illustrate fundamental concepts involving polymer physics.

2013 Annual Meeting of the American Electrophoresis Society (AES)

(392b) Dual-Electrode Electrochemical Detection for Microchip Electrophoresis: Voltammetric Identification of Chemically Labile Species

Pichetsurnthorn, P., University of Kansas

Gunasekara, D., University of Kansas

Lunte, S.,

Reactive nitrogen species such as nitric oxide ($NO^{\cdot}$) and peroxynitrite ($ONOO^{-}$) are chemically labile species that participate in oxidative stress and nitration/nitrosylation in vivo. These species have been implicated in several cardiovascular and neurodegenerative diseases. The short life time of these molecules makes them difficult to detect, often requiring indirect methods of analysis. Microchip electrophoresis coupled to amperometric detection (ME-EC) offers fast separations and sensitive detection— allowing these species to be characterized before significant degradation. Amperometric detection normally utilizes migration time to identify analytes in a sample. For complex samples such as cell lysates, analyte identification solely utilizing migration time becomes problematic when contamination protrudes. Therefore, a ME-EC method with dual electrodes was developed for identification of analytes by voltammetric characterization. Voltammetric information for analytes was obtained through a current ratio generated by employing two working electrodes in a series configuration. The current ratio can be unique to analytes with different half-wave potentials and deviations from such can imply impurities. The electrodes were integrated into a 5 cm simple "T" microchip. In this setup, the first electrode is in in-channel configuration while the second electrode is in end-channel configuration. Nitrite, tyrosine and hydrogen peroxide standards were used to optimize the system. Current ratios for these standards were generated by correcting sensitivity differences between two working electrodes. Applying this to commercially available peroxynitrite samples, it was found that test samples were contaminated with hydrogen peroxide, which is used in peroxynitrite synthesis. This method will be employed to identify RNS production in bulk cell lysates. The ultimate goal of this project is to identify the heterogeneity of reactive nitrogen species production in single cells.

(403a) Rapid Dialysis in Microfluidic Devices Using Hydrogel Membrane Micro-Windows: Phoretic Migration Under Imposed Gradients

Paustian, J. S., University of California, Santa Barbara
Squires, T. M., University of California, Santa Barbara
Originally presented on: 11/6/2013 8:30:00 - 8:51:00

We have developed a simple microfluidic technique to synthesize in-channel hydrogel structures with precise, high resolution microscale geometries. Using a microscope and UV lamp, we use projection lithography to photopolymerize thin (W=10-25μm) hydrogel membrane 'microwindows' (HMMs) into standard microfluidic devices. These microwindows are permeable to solute and solvent diffusion, yet act as rigid walls from the standpoint of fluid flow. Reservoirs of solution may thus be rapidly imposed, switched and maintained on one side of the HMM using standard microfluidic techniques, provoking changes in solution conditions on the other side without active mixing, stirring, or diluting to the sample-containing solution.

Here, we investigate steady and dynamic salt transport within HMMs. By imposing steady concentration gradients across the HMMs, comparisons of permeability properties of HMMs may be obtained. In particular, we control the effective pore size within the HMM and investigate the effect on the diffusive and Darcy permeability of the hydrogel. Dynamics of salt transport are also investigated by quickly switching reservoir solutions and visualizing the HMM and sample channel response. Fast switching is demonstrated by complete switching of salt solutions (0 to 250 mM NaCl) in a flow-free microfluidic channel in under 6 seconds. We investigate the salt transport mechanisms within the hydrogel which sets this HMM response time, and exploit this rapid transient time to investigate the phoretic motion of suspended colloids.

(403b) The Electrokinetic Properties of Cationic Surfactants Adsorbed On a Hydrophobic Substrate: Effect of Chain Length and Concentration

Azadi, G., Brown University
Vlahovska, P. M., Brown University
Tripathi, A., Brown University
Originally presented on: 11/6/2013 8:51:00 - 9:12:00

Electrokinetic(EK) properties such as the electro-osmotic flow(EOF) are influenced by surfactant adsorption at the solid-liquid interface. With the growing popularity of poly(methyl methacrylate) (PMMA) as the substrate for polymeric-based microfluidics, it is important to understand the effect of surfactants on EOF in these devices. Here, we investigate the effect of surfactant chain length and concentration on the electro-osmotic (EO) mobility induced by three cationic surfactants cetyltrimethylammonium bromide (CTAB), tertradecyltrimethylammonium bromide (TTAB), and dodecyltrimethylammonium bromide (DTAB) in PMMA microcapillaries. The presented EO measurements provide a simple and rapid method for characterization of surface hydrophobicity through adsorption of surfactants. The EO mobility as a function of concentration shows three regimes. First, at very low concentrations below 0.002mM the mobility is constant and approximately equal to the value obtained with the surfactant-free electrolyte (1mM KCl). Next, the EOF reverses and mobility increases linearly with surfactant concentration. Finally, the mobility reaches a plateau at a concentration well below critical micelle concentration (CMC), 0.2mM CTAB, 0.5mM TTAB and 2mM DTAB and decreases at the vicinity of CMC. Our results show that the rate of change in mobility with respect to concentration is a linear function of chain length and increases with longer chain surfactants. In addition, we deduce the magnitude of Van der Waals or cohesive energy between the adsorbed alkyl chains from the EO mobility values. For the alkyltrimethyl ammonium surfactants adsorbed on the hydrophobic surface of PMMA, this energy was found to be 0.114KT, smaller than the reported value for ionic surfactants adsorbed on a hydrophilic surface.

(403c) Spatial and Temporal Analysis of 2nd-Kind Electro-Osmotic Instability in Cross-Flow

Andersen, M. B., Stanford University
Druzgalski, C. L., Stanford University
Nichols, J. W., Stanford University
Mani, A., Stanford University
Originally presented on: 11/6/2013 9:12:00 - 9:33:00

For a long time, the theory of electrochemical processes was unable to explain experimentally measured electric current above the diffusion limit, the so-called overlimiting current. According to theory, the diffusion limit specifies the maximum electric current that can pass from an aqueous electrolyte into an adjacent ion-selective surface. Examples of such ion-selective surfaces include electrodes, ion-selective membranes and nanochannels, which are key components in applications such as electrodialysis, flow batteries and electrodeposition. Recently, however, 2nd-kind electro-osmotic flow and instability due to a non-equilibrium electric double layer were conclusively tied to overlimiting current. For example, in quiescent micro-channels, the formation of Rubinstein—Zaltzman vortices enhances ion transport, and creates additional electric current. Furthermore, temporal linear stability analysis has proved successful in predicting the onset and growth rate of these vortices. Similar vortices have also been observed in micro-channels with a mean cross-flow (Kwak et al., PRL, 2013), a configuration of some technological importance. In this case, however, we argue that the spatial branch of the stability problem is more relevant than the temporal branch. In particular, we apply spatial stability analysis to explore whether the system is absolutely or convectively unstable. An absolutely unstable system supports self-sustaining oscillations which can be linked to global instability. Instability in a convectively unstable system, on the other hand, is washed downstream as it grows, implying that unsteadiness must be supported by external perturbations. We consider the coupled Navier—Stokes and Poisson—Nernst—Planck equations in our model of the developing electro-diffusion layer. We apply an approximate similarity transform to describe the 2D base state in a single coordinate, which is then discretized by Chebyshev polynomials for spectral accuracy. Absolute vs. convective instability is determined by location of pinch points of the dispersion relation in the complex wavenumber plane, numerically resolved by solution of a sequence of matrix eigenvalue problems. The system is studied for different parameters including the dimensionless electrostatic screening length, the applied voltage, and the strength of the cross-flow. We also compare our results to a linear impulse response derived from a full DNS calculation. Our results are important for the understanding of overlimiting current and provide insight towards system optimization.

(403d) The Influence of Dielectric Decrement On Electrokinetics

Zhao, H., University of Nevada Las Vegas
Zhai, S.,
Originally presented on: 11/6/2013 9:33:00 - 9:54:00

We treat the dielectric decrement induced by excess ion polarization as a source of ion specificity and explore its impact on electrokinetics. We employ a modified Poisson-Nernst-Planck (PNP) equations accounting for the dielectric decrement. The dielectric decrement is determined by the excess ion polarization parameter \alpha and when \alpha=0 the standard PNP model is recovered. Our model shows that ions saturate at large zeta potentials. Because of ion saturation, a condensed counterion layer forms adjacent to the charged surface, introducing a new length scale, the thickness of the condensed layer (lc). For the electro-osmotic mobility, the dielectric decrement weakens the electro-osmotic flow owing to the decrease of the dielectric permittivity. At large zeta, when \alpha is not zero, the electro-osmotic mobility is found to be proportional to one half of the zeta potential, in contrast to zeta predicted by the standard PNP model. This is attributed to ion saturation at large zeta potentials. In terms of the electrophoretic mobility Me, we carry out both an asymptotic analysis in the thin-double-layer limit and solve the full modified PNP model to compute Me. Our analysis reveals that the impact of the dielectric decrement is intriguing. At small and moderate zeta potentials, the dielectric decrement decreases with an increasing \alpha. At large zeta potentials, it is well known that the surface conduction becomes significant and plays an important role in determining Me. It is observed that the dielectric decrement effectively reduces the surface conduction. Hence in stark contrast, Me increases as \alpha increases. Our predictions of the contrast dependence of the mobility on \alpha at different zeta potentials qualitatively agree with experimental results on the dependence of the mobility among ions and provide a possible explanation for such ion specificity. Finally, the comparisons between the thin-double-layer asymptotic analysis and the full simulations of the modified PNP model suggest that at large zeta potentials the validity of the thin-double-layer approximation is determined by lc rather than the traditional Debye length.

(403e) Tilted Post Arrays: DNA Electrophoresis in Anisotropic Media

Dorfman, K. D., University of Minnesota
Chen, Z., University of Minnesota

Originally presented on: 11/6/2013 9:54:00 - 10:15:00

Microfabricated post arrays are one microfluidic method for separating long DNA. The vast majority of experiments focus on ordered hexagonal arrays, although there are some data for square arrays, quasi-hexagonal arrays, and disordered arrays. Hexagonal arrays are anisotropic media. This anisotropy contrasts sharply with gel electrophoresis, where the random fibers of the gel are normally modeled as an isotropic medium. We will present Brownian dynamics simulation data that show that the electrophoretic mobility and dispersivity undergo a qualitative shift when the electric field vector is no longer coincident with the lattice vectors of the post array. Moreover DNA electrophoresis in such "tilted" post arrays features electrophoretic mobilities that are relatively insensitive to the electric field. As a result, operating a tilted post array at a high electric field can lead to exponential reductions in the time required to achieve a desired resolution when compared to typical post arrays. Our simulations results predict that these "tilted" post arrays provide superior separations. Although the DNA dynamics in a post array are complicated, the electrophoretic mobility results indicate that the "free path", i.e., the average distance of ballistic trajectories of point sized particles launched from random positions in the unit cell until they intersect the next post, is a useful proxy for the detailed DNA trajectories. The analysis of the free path reveals a fundamental connection between anisotropy of the medium and DNA transport therein that goes beyond simply improving the separation device.

(403f) Advancement in the Modeling of Insulator Based Dielectrophoresis

Koppula, K. S., Rochester Institute of Technology
Gencoglu, A., Rochester Institute of Technology
Olney, D., Rochester Institute of Technology
La Londe, A., Rochester Institute of Technology
Lapizco-Encinas, B. H., Rochester Institute of Technology
Originally presented on: 11/6/2013 10:15:00 - 10:36:00

Microfluidics is a new field that has gained a lot of interest in the past decade because of the potential for bioanalytical and biomedical applications. Electro-osmotic and Dielectrophoretic forces can be selectively applied to bioparticles and cells to separate or concentrate in a microfluidic device. These forces can be well manipulated in insulator-based dielectrophoretic (iDEP) devices using direct and alternating current sources with the advantage of having little or no contact of the cell or particles with the electrodes. Modeling is an essential tool to understand and tailor the dielectrophoresis systems for specific applications. The present project discusses the advancement of modeling in iDEP devices. These devices have been successfully used for the enrichment of many types of bioparticles, by trapping particles inside microchannels with arrays of insulating structures. The effect of trapped particles and their properties on the electric field distribution is studied with the aid of modeling. Factors like particle concentration and size distribution on the particle trapping is also analyzed. The study also addresses how 3D models and simulations can offer a better understanding of the volume of the particles and the trapping region and help with design improvisations.

(403g) Theoretical Investigation of Polarizability of Soft Biological Particles

Dingari, N. N., Massachusetts Institute of Technology
Buie, C. R., Massachusetts Institute of Technology
Originally presented on: 11/6/2013 10:36:00 - 10:57:00

Theoretical investigation of polarizability of soft biological particles

Naga Neehar Dingari and Cullen R. Buie

Abstract

We present a detailed theoretical model to investigate the effect of 'soft' polyelectrolyte layers on the polarization of biological particles such as bacteria. Building off of the work done by Zhao et.*al*[1], we model the soft layer of biological particles including dissociation of ionogenic charged groups and specific interactions with the background electrolyte. We model ionic transport around the soft particle with a modified Poisson-Nernst-Plank equation.[2] The fluid flow around the particle is modeled by a modified Stokes equation accounting for permeability of the soft layer. To demonstrate our model we consider two test cases: fibrillated and unfibrillated *S. salivarius* bacteria whose electrophoretic mobilities were analyzed theoretically and experimentally by Duval et.*al*[3].We explore a wide range of bulk electrolyte concentrations to consider both thin and thick double layer limits. The results demonstrate an interesting interplay between soft layer conductivity and double layer conductivity on the polarizability of these two strains of *S. salivarius*. The soft layer conductivity is a function of pH, pKa of ionogenic groups and bulk electrolyte concentration whereas double layer conductivity is a function of bulk electrolyte concentration. Under low concentration regimes, double layer conductivity dominates over soft layer conductivity and these two strains exhibit widely different dipole moment coefficients. The dipole moment coefficients are also stronger functions of pH at low bulk electrolyte concentrations whereas they approach a constant value of -0.5 at higher concentrations. As a consequence of our analysis, we highlight electrolytic and pH conditions favorable for manipulation of biological particles using dielectrophoresis and electrophoresis depending on the nature of their soft layers.

(1) Uppapalli, S.; Zhao, H. *Langmuir□: the ACS journal of surfaces and colloids* **2012**, *28*, 11164–72.

(2) Duval, J. F. L.; Ohshima, H. *Langmuir□: the ACS journal of surfaces and colloids* **2006**, *22*, 3533–46.

(3) W., Norde; Duval, J. F. L *Langmuir: the ACS journal of surfaces and colloids* **2005**, *21*, 11268.

(481a) Investigation of a Novel Platform for Manipulation of Microparticles Using Dielectrophoresis

Angione, S., Brown University
Croote, D., Brown University
Karlberg, S.,
Tripathi, A., Brown University
Originally presented on: 11/6/2013 12:30:00 - 12:51:00

Investigation of a Novel Platform for Manipulation of Microparticles using Dielectrophoresis

Stephanie Angione[1], Derek Croote[1], Sara Karlberg[2], Anubhav Tripathi[1]

[1]Center for Biomedical Engineering, School of Engineering and Division of Biology and Medicine, Brown University, Providence, RI

[2] School of Biotechnology, KTH Royal Institute of Technology, Stockholm, Sweden

Introduction:

Dielectrophoresis (DEP) is a commonly used tool for the study and manipulation of micron and sub- micron particles, including biological molecules, in solution. Because biological molecules of interest are often found at concentrations so low that immediate analysis by current methods is not possible, pre-concentration is necessary as a prior step. The field of diagnostics is plagued by the use of off-chip pre-concentration methods, which prevents real-time testing in resource-poor settings at the point of care. Current DEP platforms do not address the needs of diagnostic applications where pre-concentration of sample must be completed in a rapid and simple manner. Our novel platform, is a potential solution to this problem as it combines the established science of DEP for the manipulation of DNA, proteins and cells with a unique electrode material and geometry.

Materials and Methods:

Polystyrene microparticles (FluoSpheres, Invitrogen) were characterized by the manufacturer as being 9.9 Â±0.1188 Âµm in diameter, with a density of 1060 kg/m^3. The stock concentration was given as 3.6 x 10^6 beads/mL, and further diluted in 1000 Âµl of the appropriate conductivity KCl solution. Experiments were performed on a Nikon Eclipse TE2000 inverted microscope equipped with a Pixelink PL-B681 CMOS camera for image and video acquisition. The polydimethylsiloxane (PDMS) microchip containing micro-channels and reservoirs was designed to allow for imbedding of the pipette tips within the design. PDMS microchip fabrication was done using standard photolithography procedures.

Results and Discussion:

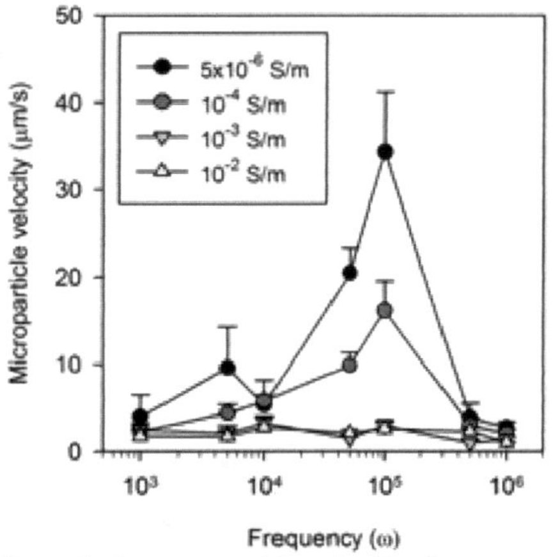

Figure 1: Average particle velocities between conductive pipette tips as a function of medium conductivity and applied electric field frequency. Data are presented mean ±SD,n=3.

We present the validation of an innovative electrode material and PDMS microchannel geometry for dielectrophoretic concentration of 10 Âµm polystyrene beads. We have implemented carbon black compounded polypropylene pipette tips as viable electrodes for the generation of electric fields in solution that can be implemented to concentrate and target particles within the pipette tips for easy removal. We have performed modeling and experimental validation of the particle movement, which indicates that microparticle behavior is a function of conductivity and frequency. The observed relationship of particle velocity to medium conductivity and applied electric field frequency has practical implications for particle manipulation using conductive pipette tips. To accomplish successful trapping, the particle velocities were maximized and Figure 1 displays the microparticle velocity as a function of frequency and conductivity, indicating that the $5x10^{-6}$ S/m conductivity solution with an applied electric field frequency of 10^5 Hz was optimal for increasing the particle motion. This platform provides an innovative way to improve diagnostic techniques and sensitivity by targeting, trapping and concentrating particles.

Conclusion:

The aim of this work was to create a platform to utilize and validate carbon black compounded polypropylene pipette tips as viable electrodes for the generation of electric fields in solution. The use of this innovative electrode material for dielectrophoresis of 10 Âµm polystyrene particles was established as a proof of concept for pre-concentration, trapping and targeting of particles. We characterized the behavior of microparticles in solutions of varying conductivities under the effects of applied electric fields of varying frequencies to optimize particle motion and trapping. Further work will focus on concentrating and trapping DNA from dilute solutions for direct amplification and detection using nucleic acid amplification methods.

(481b) Design of Electrothermal Pumps Using Resistive Heaters

Williams, S. J., University of Louisville
Green, N. G., University of Southampton
Originally presented on: 11/6/2013 12:51:00 - 13:12:00

Electrothermal pumping induces fluid motion through the interaction of an electric field with temperature gradients within the fluid. Non-uniform heating is traditionally generated through Joule heating. Therefore, the applied electric field serves two purposes simultaneously: (i) to heat the fluid through internal Joule heating and (ii) to drive electrothermal motion. This presentation describes a new method of electrothermal pumping using thin film resistive heaters. External heating, in effect, introduces non-uniform temperatures independent of the applied field. Various external heating design configurations will be introduced and compared to internal heating mechanisms. Geometries include (i) asymmetric electrodes, (ii) symmetric interdigitated electrodes, and (iii) travelling wave electrothermal pumping. Numerical simulations will demonstrate the advantages of using thin film resistive heaters. For example, using asymmetric electrodes, electrothermal pumping using thin film heaters is over 2.5 times more energy efficient than Joule heating alone. In addition, external heating allows for electrothermal pumping to be applied towards low conductivity samples.

(481c) Design and Simulation of An Automated Rare Blood Cell Detector

Qian, Z., Auburn University

Boland, E., Techshot Inc

Todd, P. W., Space Hardware Optimization Technology, Inc

Hanley, T. R., Auburn University

Originally presented on: 11/6/2013 13:12:00 - 13:33:00

Design and Simulation of an Automated Rare Blood Cell Detector

Zhixi Qian[a], Eugene D. Boland[b], Paul W. Todd[b]. Thomas R. Hanley[a],

[a]Department of Chemical Engineering, Auburn University, AL 36849

[b]Techshot, Inc. 7200 Highway 150, Greenville, IN 47124

In recent years the use of circulating tumor cell (CTC) counts in cancer management has increased. Detecting small numbers of rare cells in blood and non-invasively and monitoring the therapeutic progress, CTC detection technology currently plays an important role. However, the detection can be performed only in specialized laboratories, and the analysis requires several skilled technicians, making use expensive. Also, results require weeks (or even months), making it difficult to detect disease at an early stage.

The primary objective of this research is to design, simulate and test an accurate, cost-effective, user-friendly point-of-care CTC detection device. The preliminary design includes three syringes for sample, buffer and reagents, a static mixer, a magnetic filter, a microfluidic "chip" and a waste receptor. The device is smaller and cheaper than traditional detection devices. Also, device operation is easier with only one operator required. The device can be used in clinics, nursing homes or residences allowing earlier detection of most fatal cancers, including pancreatic cancer, ovarian cancer and melanoma.

Our recent research focuses on the analysis and simulation of static mixer and the flow cytometer as well as further analysis of factors affects the fluid's motion pattern. The fluid dynamic analysis of the staged mixer was accomplished using the FLUENTTM analysis package, modeling of fluid flow and tracking particles' movement in complex geometries. The simulations include the effects of the magnetophoretic mobility and gravity as well as the effect of hinder flow. Results from experimental separations of CTC's from blood samples are compared to simulation results.

Marshall, L. A., Stanford University

Rogacs, A., Stanford University

Meinhart, C., UCSB

Santiago, J. G., Stanford University

Originally presented on: 11/6/2013 13:33:00 - 13:54:00

(481d) An Injection-Molded Device for Purification of Nucleic Acids from Whole Blood using Isotachophoresis

Lewis A. Marshall, Anita Rogacs, Carl Meinhart, and Juan G. Santiago

Stanford University

Stanford, CA 94305, USA

We have designed a new microfluidic chip to simplify nucleic acid preparation from biological samples such as whole blood. We previously demonstrated the application of isotachophoresis (ITP) for sample preparation.[1] We here describe the design of an injection molded plastic device capable of purifying DNA from a 25 µL sample of whole blood lysate under 40 min.

We have targeted samples with volumes between 2-25 µL in an attempt to address an unmet need in sample preparation. Solid phase extraction (SPE) kits for purification of nucleic acids work well for sample volumes of 100-1000 µL, and nucleic acid yields of several nanograms. However, they exhibit poor extraction efficiency for samples of low volume and smaller nucleic acid yields. ITP does not rely on surface adsorption, and so can operate on samples with a wide range of volumes and nucleic acid content. Our new device enables ITP purification of samples in this critical volume range while simplifying device operation. We have also maximized device throughput by adjusting our chemistry and protocol and considering the sample volume processing rate as a cost function..

This new device is pictured in **Figure 1**. This custom device is fabricated from injection molded Topas cyclic olefin copolymer (COC) or poly(methyl methacrylate) (PMMA). The device was fabricated by the microfluidics foundry Microfluidic Chipshop GmbH of Jenna, Germany. The injection molded layer contains channels with a nominal cross-section of 2 mm x 150 µm, a total length of 16 cm. The injection molded layer is capped by a thin, transparent plastic film to enable imaging into the channel.

We've included novel features in this chip to improve the device loading process. To simplify set up of the initial interface between the sample and the separation buffer, we've developed a specialized capillary barrier structure that provides passive valving for sequential loading of liquids into the device. This structure is shown in **Figure 2**. In addition to incorporating this passive valving system for device loading, we have incorporated the use of gels in the buffering reservoirs to suppress pressure driven flow. This simplifies loading by making the device performance robust to uneven fluid levels in the reservoirs.

We used this device to extract DNA from whole blood lysate samples. A visualization of the separation process is shown in **Figure 3**. Here, we separate two fluorescent dyes with mobilities similar to DNA and a contaminating protein species. The analyte dye focuses into a sharp band and elutes into the extraction reservoir, while the contaminating species lags behind and within the channel for later disposal.

We verified the purity of our extracted samples from species that inhibit polymerase chain reaction (PCR) by performing quantitative PCR (qPCR) and comparing the results with those collected from samples extracted using a commercial SPE DNA purification kit. Sample qPCR data is shown in **Figure 4**.

By introducing a standard injection molded device for purification of samples using ITP, we have created a method of performing sample preparation at a relevant scale using ITP. This system addresses an unmet need in sample preparation of nucleic acids by providing an alternative method for processing samples with low nucleic acid content.

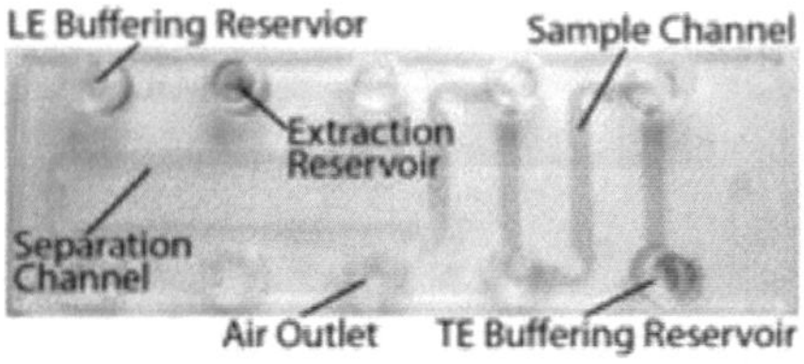

Figure 1. Image of the new injection molded device. For imaging, the device was filled with colored liquid to improve contrast. The sample channel (red) can hold a 25 μL sample. The separation channel (blue) holds 28 μL separation buffer. There are buffering reservoirs for both the leading and trailing electrolyte (LE and TE). These buffering reservoirs contain a gel with high buffering capacity gel to maintain a constant pH in the channel while suppressing pressure-driven flow. An air outlet with specialized capillary structures is included to ease the loading process, as shown in **Figure 2**.

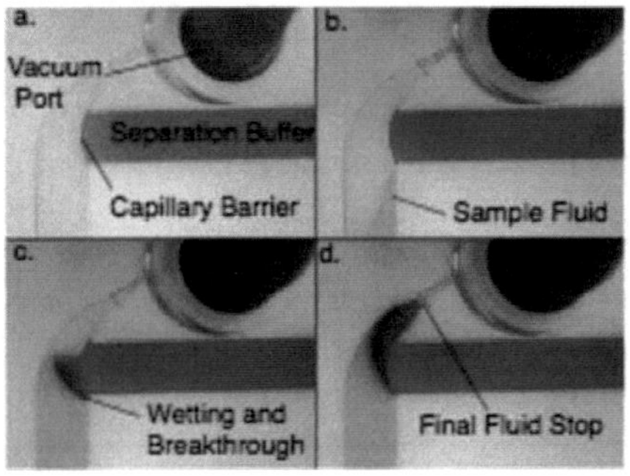

Figure 2. Capillary loading structures near the vacuum port. The capillary structures allow set-up of a fluid interface without wasting sample or trapping bubbles. a. The separation channel is initially filled with the (red) separation buffer, which wets up to the surface of the capillary barrier. b. The sample is applied to the sample channel, and a -0.1 psig vacuum is applied to the air outlet to draw the sample into the channel. c. The sample wets the separation buffer, causing it to break through the barrier. d. The liquids form a sharp interface, and flow up to the final fluid stop barrier.

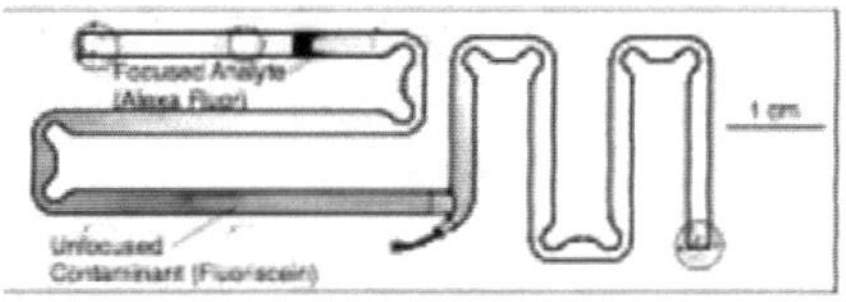

Figure 3. An experimental image of separation of two dyes overlaid with the device schematic. Two dyes, Alexa Fluor 488 (AF) and fluorescein (FL) are loaded as the sample. When current is applied, the AF focuses to the ITP interface, while the FL remains behind as an unfocused zone. The AF is about to elute into the extraction reservoir, while the FL remains behind in the channel.

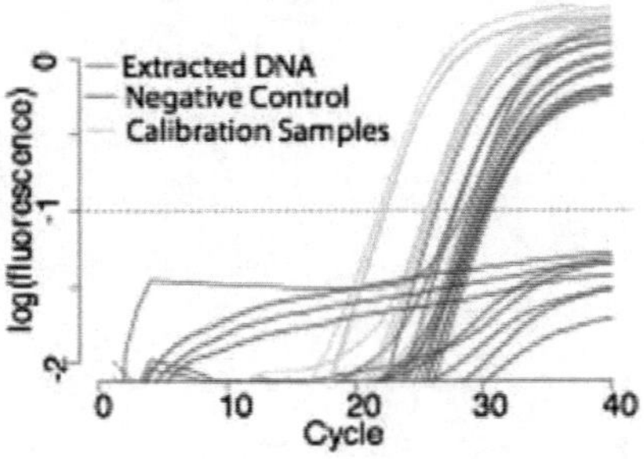

Figure 4. qPCR results. We performed qPCR with samples extracted from whole blood using isotachophoresis. The extracted DNA amplified consistently, and at the predicted temperature for the amplified sequence (not shown). The results were comparable to qPCR amplification of calibration samples with template DNA extracted from blood using a commercially available SPE kit. We performed negative controls using whole blood lysate, water, and the separation buffer as templates, and observed no amplification.

REFERENCES:

1. "Purification of nucleic acids from whole blood using isotachophoresis," Persat, A., Marshall, L. A., and Santiago, J. G. Analytical Chemistry **81**, 9507 (2009).

(481e) Microfluidic Electrokinetic Sample Holder for Serial Femtosecond Crystallography

Sierra, R., SLAC National Accelerator Laboratory
Laksmono, H., SLAC National Accelerator Laboratory
Bogan, M., SLAC National Accelerator Laboratory
Originally presented on: 11/6/2013 13:54:00 - 14:15:00

We present a microfluidic electrokinetic sample holder (MESH) which delivers suspensions of protein crystals to a hard X-ray free electron laser (XFEL) source at the Linac Coherent Light Source (LCLS) at SLAC National Accelerator Lab, for structural determination by serial femtosecond crystallography (SFX). The MESH allows for low sample consumption rates on the order of hundreds of nanoliters per minute of protein suspension; 10x to 100x lower than current sample delivery technologies, such as hydrodynamically focused jets and aerodynamically focused particle streams. The applied electric field and added viscosity, necessary for operating the MESH, diminish the background signal and reduce the flowrate, while not affecting the structure of protein crystals as observable down to better than 4 Angstroms. Simple, off-the-shelf components combined with small sample consumption rates on the order of micrograms, make MESH an ideal sample delivery method for protein structural determination at hard X-ray sources, from simple protein standards like thermolysin, to complex membrane proteins like photosystem II.

simulations were used to create a time-dependent 'pulse envelope' that would be experienced by a cell experimentally. A three dimensional model of a single cell including internal organelles was constructed and exposed to this pulse envelope. The resulting increase in transmembrane potential was then calculated and compared literature values for cellular response and damage.

Master stamps of the two theoretical devices were created on separate silicon wafers using deep reactive ion etching. Polymer replicates were then cast in PDMS with a 10:1 ratio of polymer to curing agent. The polymer replicates were bonded to glass slides after exposure to air plasma for two minutes and stored under vacuum. Prior to experimentation, the fluid electrode channels were filled with phosphate buffered saline with a conductivity of 1.4 S/m. Cells were suspended in a sucrose solution with a conductivity of 0.01 S/m and driven through the devices using a syringe pump. A high voltage amplifier and series transformer were used to apply voltages with amplitudes up to 300 V_{RMS} between 10 and 800 kHz.

The amplitude of the applied voltage and the fluid velocity was adjusted to determine the experimental doses required to induce reversible and irreversible electroporation effects. This work confirms the existence of a relationship between the Clausius-Mossotti factor and the ability to induce electroporation effects. The energy required to electroporate cells was found to increase with as the frequency of the applied field was elevated. The contactless method presented here may be advantageous as it eliminates any possible electro-chemical effects or bubble formation at the fluid-electrode interface. Future work will focus on evaluating cells off chip for transfection efficiency.

2013 Annual Meeting of the American Electrophoresis Society (AES)

(481f) Semi-Preparative Isotachoporesis for Fractionation of RNA From Blood

Meagher, R. J., Sandia National Laboratories

Eid, C.,

Originally presented on: 11/6/2013 14:15:00 - 14:36:00

RNA transcripts circulating in blood or blood plasma may be useful as biomarkers of infection, host response to infection, or numerous other states of health. RNA-Seq technologies offer the ability for quantitative, unbiased analysis of all circulating RNA within a clinical sample in a single experiment, and development of sample preparation techniques for RNA from blood is an area of active investigation. Circulating RNA tend to heavily degraded, and are typically <200 nt in length. This pool includes both "small" RNA (microRNA, etc), as well as degraded mRNA and other larger species. Both pools can be informative, but are ideally analyzed separately due to disparate requirements for library preparation, sequencing techniques, and bioinformatics analysis. We have thus developed a technique for small-scale preparative fractionation of RNA into "small" and "large" fractions, utilizing discontinous zone electrophoresis in a "large-channel microfluidics" device sized to process tens to hundreds of nanograms of RNA, acheiving good recovery from samples at the scale of tens of microliters. Our device operates at a larger scale than typical analytical separations, but at a smaller scale than commercial kits and devices that are mainly geared to fractionating DNA at microgram or larger scale. We discuss design of the device and effect of operating parameters such as buffer system on performance of the device, and discuss constraints for use in the workflow of preparing sequencing libraries.

Originally presented on: 11/6/2013 14:36:00 - 14:57:00

AIChE 2013

Topical 3: 2013Annual Meeting of the American Electrophoresis Society

?T3006 ? ?Electrokinetics for Sample preparation?(

(481g) DEP Isolation of Cancer Related Circulating Cell Free (CCF) DNA Biomarkers Directly from Blood

Michael J. Heller[1,2,*]Avery Sonnenberg[1], Jennifer Y. Marciniak[1], Laura Rassenti[3], Emanuela Ghia[3], Elaine Skowronski[1], Sareh Manouchehri[1], George Widhopf[3] and Thomas J. Kipps[3]

University of California San Diego, Department of Bioengineering[1], Department of Nanoengineering[2], and UCSD Moores Cancer Center[3], La Jolla, CA, USA 92093-0448

Abstract ? Circulating cell free (ccf) DNA/RNA biomarkers are now widely used for cancer diagnostics and management. Unfortunately, the isolation of ccf-DNA/RNA directly from blood has not been possible, and its isolation from plasma is a long and involved process. We report on an AC electrokinetic approach for the rapid isolation of ccf-DNA directly from a small volume of un-processed blood. A dielectrophoretic (DEP) microarray device (chip) was used to isolate ccf-DNA directly from fresh blood samples of fifteen Chronic Lymphocytic Leukemia (CLL) patients and three healthy individuals. Ccf-DNA/RNA from 25uls (a drop) of blood was separated and concentrated into DEP high-field regions in about three minutes, while blood cells, proteins and other biomolecules were removed by a fluidic wash. Concentrated ccf-DNA (SYBR-Green stained) was detected by fluorescence and then eluted for PCR and DNA sequencing. The complete process, blood sample to ccf-DNA for PCR, was completed in less than 10 minutes. Eluted ccf-DNA, from 5ul of the original CLL blood sample, was amplified by PCR using IGHV-specific primers to identify the unique IGHV gene expressed by the leukemic B-cell clone (ie the major IGHV subgroups expressed by CLL B cells IGHV1, IGHV3, IGHV4). The PCR results obtained by DEP from CLL blood were comparable to results obtained using conventional sample preparation of ccf-DNA which starts with one ml of plasma. The DNA sequencing results from the PCR amplified cff-DNA obtained by DEP correlated with earlier sequencing results from genomic DNA isolated from the CLL patient's leukemic B cells. The ability of DEP to provide rapid isolation and detection of ccf-DNA from the equivalent of a drop of blood represents a major step forward in the quest for ?liquid biopsy? and point of care (POC) cancer diagnostics and patient monitoring.Â Â Â

Introduction - AC electrokinetic methods like dielectrophoresis (DEP) are well-known techniques for achieving separations of cells, nanoparticles, and biomolecules. However, until recently DEP technology had remained impractical for general use with biological/clinical samples (blood, plasma, serum) and in high conductance buffer solutions (>1mS/cm). For example, earlier DEP work on separating bacteria from blood (~7-9 mS/cm) required a 50-fold dilution of the blood sample before the DEP separation process could be carried out [1]. More recently, we have developed new DEP devices and methods that allow hmw-DNA, nanoparticles, bacteriophage and circulating cell free (ccf) DNA to be isolated and detected from both high ionic strength buffers and blood [2-6]. Using special microarrays with robust platinum microelectrodes over-coated with novel hydrogel layers (designed and fabricated by Biological Dynamics, La Jolla, CA) we are now able to demonstrate the rapid isolation, fluorescent detection and subsequent PCR amplification and DNA sequencing of cancer related ccf-DNA directly from a 25 ul sample of un-processed blood from Chronic Lymphocytic Leukemia (CLL) patients.

Results: Figure 1 (below) shows the electrokinetic DEP device (Biological Dynamics) used for the separation of ccf-DNA from CLL patient blood. The DEP microarray device/chip is a 10 x 12 mm silicon die patterned structure with an interdigitated array of 1000 (60 Âµm dia.) platinum microelectrodes on 160 um center-center pitch. The microarray is over-coated with a 200nm-500nm porous hydrogel layer. A printed circuit board is attached to the chip using electrically conductive pressure sensitive adhesive which provides electrical contact to standard pogo pin connectors. The PCB cutout forms the sides of the flow cell and an acrylic window with fluidic input and output ports forms the top cover, resulting in a 20 ul flow cell volume. Typically, ~25 ul of the sample would be aspirated to completely fill the flow cell and the fluidic ports. For the actual process, the chip was first pre-treated with approximately 25 ul of 0.5X PBS and run at 2Vrms, 5Hz sine waveform for 15 second to improve the hydrogel porosity. The fluid was aspirated and 25 ul of undiluted sample (blood) was then applied to the chip flow cell and run at 12 Volts peak to peak (Vp-p) at 10 kHz (sine wave) for 3 minutes.

Figure 1 ? Electrokinetic DEP microarray chip device designed and fabricated by Biological Dynamics for carrying out separations in high conductance buffers and clinical samples (blood, plasma, serum, etc.)

A checker board field geometry is formed by the AC bias with the positive DEP high-field regions occurring on the microelectrodes, and negative DEP low-field regions occurring between microelectrodes. With the AC voltage ON, the flow cell was washed with TE buffer for 5 minutes at 100 ul/min flow rate. After the wash step, the power was turned OFF allowing the captured DNA to diffuse off the hydrogel. The sample was then aspirated from the flow cell and stored at 4° C in a microcentrifuge tube. In order to visualize real time capture on the microelectrode array, the CLL and normal blood samples were treated with 5x SYBR Green I fluorescent dye which stains the double stranded DNA. Bright field and fluorescent images of the microelectrode pads were then acquired using a CCD camera with a 10x objective, FITC fluorescent filters, and a 470nm LED excitation source. Figure 2 shows representative fluorescent images of the SYBR Green stained ccf-DNA from one of the normal blood samples along with two CLL patient blood samples. MATLAB was used to create 3D fluorescent intensity images which provide better quantification of the relative amount of ccf-DNA that has been concentrated onto the microelectrodes by the DEP process. In general, fluorescent DNA levels were higher in all 15 CLL patient samples than in the normal blood samples; and in most cases they were significantly higher. The concentration of ccf-DNA in the eluted samples (25ul) was also determined by PicoGreen fluorescent analysis. Again the fluorescent DNA levels were higher in all 15 CLL patient samples than in the normal blood samples, and in most cases the concentrations were significantly higher. Â Subsequent PCR amplification for ccf-DNA (equivalent to 5ul blood) isolated by DEP from all fifteen CLL blood samples compared well with the results obtained from ccf-DNA isolated by the Qiagen process starting with 1ml of CLL patient plasma. Finally, DNA sequencing carried out on the PCR amplified ccf-DNA isolated by DEP correlated very well for all 15 CLL patient samples with sequencing results (Gold Standard) from genomic DNA obtained from the patients B-lymphocyte cells which was isolated from 5 ml of blood.Â

Figure 2 - Fluorescent images of the SYBR green stained ccf-DNA from a normal blood sample and two CLL patient blood samples. Also shown (far left) are 3D fluorescent intensity images created by MATLAB.

Discussion and Conclusions

The use of circulating cell-free (ccf) DNA biomarkers for ?liquid biopsy? will unquestionably be important for future point of care (POC) cancer diagnostics and management. Unfortunately, the time and process complexity required for the isolation of ccf-DNA from plasma by conventional methods is a major limitation for developing POC diagnostic formats. Additionally, these processes add significant cost to the assays which will also limit their use. We have now shown that an AC electrokinetic DEP microarray device can be used to isolate ccf-DNA directly from CLL patient blood in less than 10 minutes. The DEP process is not only significantly faster than the conventional methods, but requires only three steps. We have also shown that DEP allows ccf-DNA to be isolated from a relatively small sample volume of only 25 uls, which is equivalent to a drop of blood.

References

(1) Cheng J, Sheldon EL, Wu L, Uribe A, Gerrue LO, Carrino J, Heller MJ and O'Connell JP, ?Preparation and Hybridization Analysis of DNA/RNA from E. coli on Microfabricated Bioelectronic Chips?, Nature Biotechnology, Vol. 16, pp. 541-546, 1998.

(2) Krishnan R., Sullivan B.D., Mifflin R.L., Esener S.C., & Heller M.J., AC electrokinetic separation & detection of DNA nanoparticles in high-conductance solutions, *Electrophoresis* **29**, 1765-1774, 2008

(3) Â Krishnan R. & Heller M.J., AC electrokinetic method for enhanced detection of DNA nanoparticles, *J. Biophotonics* **2**, 253-261**,** 2009

(4) Krishnan R., Dehlinger D.A., Gemmen G.J., Mifflin R.L., Esener S.C., & Heller M.J., Interaction of nanoparticles at the DEP microelectrode interface, *Electrochem. Comm.* **11**, 1661-1666, 2009

(5) Sonnenberg, A, Marciniak, JY, Krishnan, R, Heller MJ, ?Dielectrophoretic Isolation of DNA and Nanoparticles form Blood?, *Electrophoresis* V33, 2482-2490, 2012

(6) Sonnenberg, A, Marciniak JY, McCanna J, Krishnan R, Rassenti L, Kipps TJ and Heller MJ ?Dielectrophoretic Isolation and Detection of cfc-DNA Nanoparticulate Biomarkers and Virus form Blood?, Electrophoresis, V34, pp.1076-1084, 2013

(482a) Characterization of the Permeability of the Brain Endothelium Due to Electroporation Using a Dynamic Microengineered Model

Bonakdar, M., Virginia Tech
Garcia, P.,
Davalos, R., Virginia Polytechnic Institute and State University
Originally presented on: 11/6/2013 12:30:00 - 12:52:00

Abstract

The microvascular endothelial cells, which cover the interior of the brain microvascular network, play a significant role in controlling the transport of different molecules into the central nervous system (CNS). Although crucial to human health by maintaining the neuroparenchymal microenvironment and protecting the neural tissue from toxins, the blood-brain barrier (BBB) also limits the success rate of new therapies. One significant obstacle to drug development for targeting the CNS has been the high resistivity of the BBB in allowing large molecules to penetrate into the CNS. Electroporation, which is the permeabilization of the cell membrane using pulsed electric fields, is a non-viral physical method for drug delivery [1]. We fabricated a microengineered BBB model and evaluated its permeability to different chemicals via delivery of pulsed electric fields by using a novel laminated microdevice with embedded electrodes which accommodates the BBB mimic. Our results support the hypothesis that electroporation is able to increase the permeability of the BBB without compromising cell viability in an *in vitro* model.

Introduction

Organs on a chip, which are a class of microdevices that feature a physiologically relevant function, are a method of in vitro tissue analysis. The BBB has a significant role in screening certain molecules from the CNS and is a major obstacle to drug development for many CNS diseases. Hence, many researches are focused on studying the physiology of this barrier and how to permeate through it. Booth *et al.* [2] developed a microfluidic BBB model incorporating a co-culture of endothelial and astrocytes and monitored the formation of tight junctions using transendothelial electrical resistance. In another study Prabhakarpandian *et al.* [3] modeled a micro BBB using brain endothelial cells in astrocyte conditioned medium and investigated the effect of astrocytes on the formation of the tight junctions and the transport of certain chemicals across the barrier. Flow induced shear force and chemical factors are also deployed to modulate the permeability of the BBB [4]. In this study, we stimulated the microengineered BBB model by short pulsed electric fields (BTX ECM 830, Harvard Apparatus)) to induced cell electroporation and disrupt the blood-brain barrier. The transport of chemicals across the BBB was quantified as a function of molecular weight, electric field and flow rate.

(482b) Electroporation of Cells On Chip Using High Frequency Electric Fields Without Electrode-Sample Contact

Sano, M. B., Virginia Polytechnic Institute and State University
Davalos, R. V., Virginia Polytechnic Institute and State University
Originally presented on: 11/6/2013 12:52:00 - 13:14:00

Contactless Dielectrophoresis (cDEP) is a relatively new technique for investigating, isolating, and enriching mammalian cells. This technique separates cells from contact with the electrodes by using fluid electrodes which are isolated from the sample channel by thin insulating membranes. This technique has been used to isolate live mammalian cells from dead cells and beads of the same size as well as discriminate between cells with different metastatic behaviors. A significant effort has been made to expand the originally limited frequency spectrum over which these devices are effective. Current designs are capable of manipulating cells at frequencies between 1 kHz and 1 MHz in low conductivity, physiologically suitable buffers. This wide range of operation allows investigators to manipulate mammalian cells at frequencies close to their first Clausius-Mossotti factor crossover point where the difference between similar cells is the most pronounced. This has enabled the use of cDEP devices for a number of biological applications which analyze and culture the cells off chip, post sorting. Here, we show that these devices can additionally be used to transport large molecules or genes into cells with high efficiency while maintaining a sterile environment.

The application of an external electric field causes a redistribution of charge within a cell resulting in an increase in the cell's transmembrane potential (TMP). This, in turn, increases the free energy in the molecules in the cell membrane and the rate of transport of larger molecules into the cell. The process, known as electropermeabilization or electroporation, has been widely used in biology to transfer macromolecules, into cells while maintaining cell viability. Irreversible electroporation (IRE) results if the external field intensity increases past a certain threshold, such that the cell membrane is permanently destabilized and the cell cannot recover. Charge redistribution in presence of an electric field is not instantaneous. For a brief time, ions internal and external to the cell rearrange in response to the external field. This displacement current halts as the TMP reaches its maximum, after approximately 1 μs. If the electric field is pulsed or due to a high frequency AC voltage, a displacement current propagates through the cell giving rise to increased membrane potentials in the nucleus and organelles. In typical reversible and irreversible electroporation protocols the pulses are a single polarity with on times on the order of 100 to 1000 microseconds. In this study, we show that fields with frequency components above 10 kHz can result transient and irreversible electroporation effects.

Multi-physics models of cDEP devices were constructed in COMSOL which incorporated fluid- and electro-dynamics. The device contained constricting features in which cells are exposed briefly to an elevated electric field. By altering the magnitude of the applied voltage and the fluid velocity, the field-time dose experienced by the cells can be adjusted. Because the field varies both spatially and temporally, the

Materials and Methods

The microfluidic device was fabricated using photolithography and replication molding. A polycarbonate membrane was embedded into the device separating an upper and lower chamber (100Âµm deep and 300 Âµm wide). Rat brain endothelial cells (RBE4) were cultured on the membrane until the formation of a confluent monolayer and the expression of the tight junctions. The microdevice features a pair of microelectrodes on one side of the barrier. By applying electric pulses of different amplitudes, the endothelial cells were reversibly electroporated and the permeability of the barrier was changed. We injected Dextran (Sigma) of different molecular weights into one chamber and measured the diffused amount into the other chamber following electroporation treatments.

Results and Discussion

Brain endothelial cells were cultured on the porous membrane inside the microenvironment until a confluent monolayer of cells was formed. We repeated the experiment with the presence of astrocytes on the opposite side of the membrane. Astrocytes are star shaped glial cells in the brain and spinal cord which support the endothelial cells and contribute to the formation of the tight junctions. To have a better representation of the in vivo conditions, media was perfused over the endothelial cells in the top chamber. Figure 1 shows a schematic of the experimental setup. By applying electric pulses between the electrodes, the permeability of the barrier was temporarily increased, allowing certain molecules of high molecular weight to penetrate through the BBB. The permeability analysis after the electroporation and live/dead assay confirms the reversibility of the process without compromising cell viability.

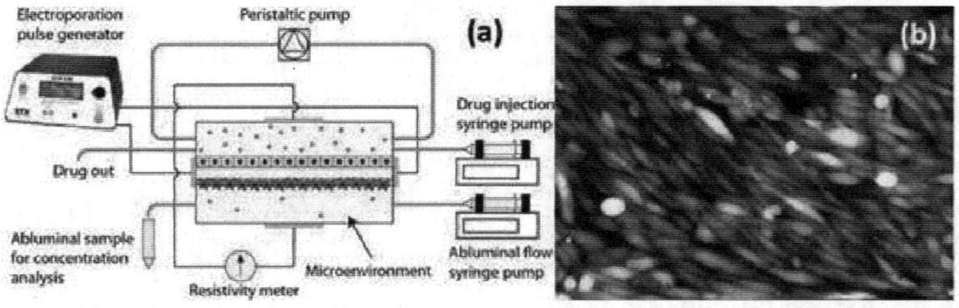

Figure 1. (a) Test setup including the BBB mimic, electrical, and flow circuits. Two halves of the microfluidics device fabricated in glass and PDMS are bonded together, encapsulating the membrane and forming separate luminal and abluminal channels Â in both sides of the membrane. (b) a monolayer of RBE4 cells cultured in the microchannel of the device.

Conclusions:

We modeled the BBB inside the microfluidic device and stimulated it by electroporation pulses to increase its permeability to certain chemicals without affecting cell viability. Using this method we are able to quantify the required electric field for delivery of certain drugs in a dynamic environment.

References:

[1] Garcia, P. et al, PLoS ONE, 2012, 7(11), e50482

[2] Booth, R. et al, Lab Chip, 2012, 12, 1784.

[3] Prabhak, B. et al, Lab Chip, 2013, 13, 1093.

[4] Griep, L. Biomed. Microdevices, 2013, 15, 145-150.

(482c) Aunps-Polyplex-Electroporation Enhanced DNA and RNA Delivery

Huang, S., Louisiana Tech University
Zu, Y., Louisiana Tech University
Lu, Y.,
Wang, S., Louisiana Tech University
Originally presented on: 11/6/2013 13:14:00 - 13:36:00

Non-viral methods have been explored as replacements of viral systems for DNA or RNA delivery to mammalian cells for their low toxicity and immunogenicity. In nonviral systems, chemical methods, including lipoplex and polyplex, exhibit advantages on probe carry, protection, and condenation while physical methods, such as biolistics and electroporation, facilitate fast, directial, and efficient probe delivery. Great improvements have been achieved on each approach over the years. However, they have yet to reach levels competitive to their viral counterparts on the delivery efficiency, cell viability, and targeting specificity. We hypothesize that the combination of some physical and chemical methods could improve the performance of nonviral delivery systems. Specifically, we immobilized polyplex (a chemical approach) on gold nanoparticles/nanorods and delivery them into cells through electroporation (a physical approach). Conjugating with AuNPs helps minimize the cytotoxicity concerns from polyplex after cytoplasmic release while still retain good probe protection in serum. Combining with electroporation, conjugated polyplex (AuNPs-Polyplex) showed quick delivery and significant enhancement on the transfection efficiency whil no obvious increase of toxicity. Specifically, gold nanoparticles (AuNPs) of various sizes and geometries were conjugated with polyplex of polyethylenimine (PEI)-plasmid and PEI-siRNA. The conjugated nanoparticles were then mixed with cells and electroporated using a commercial instrument and a serpentine flow-through electroporation (SFE) microsystem. These nanoparticles help enhance gene delivery to mammalian cells with better cellular uptake of molecular probes and reduced adverse effects associated with nonviral delivery approaches. Such a combination of physical and chemical delivery concept may stimulate further exploration in the delivery of various therapeutic materials for both *in vitro* and *in vivo* applications.

(482d) Cell Allignement Under Unidirectional Electropulsation In A Microfluidic Device

Loufakis, D. N., Virginia Tech
Cao, Z., Virginia Tech
Ma, S.,
Mittelman, D., Virginia Tech
Lu, C., Virginia Tech
Originally presented on: 11/6/2013 13:36:00 - 13:58:00

The movement of cells under an electric field is widely practiced in microfluidics. The specific characteristics of the cell trajectory are strongly affected by the properties of the electric field (EF) and the cell surface. Cell surface is covered by glycoconjugates that may be potentially charged depending on the pH. Cells that are exposed to a unidirectional EF typically move in one direction due to electrophoresis and electroosmotic flow. In the report, we demonstrate a microfluidic device, in which unidirectional, homogeneous dc pulses drive eukaryotic cells to align at the center of a sealed chamber. Cells were scattered in a microfluidic chamber which had two microfabricated gold electrodes positioned at the two ends of the chamber. The chamber was sealed up during the period when multiple dc pulses of the same direction were applied. We observed interesting movement of cells at various parts inside the chamber toward the center. We postulate that electrochemical effects dominate the tiny space and the generated pH gradient determines the cell surface charge at a local spot and produces the surprising concentration effect. This novel manipulation of a low number of cells in a microfluidic environment may have important applications to a number of cellular studies.

(482e) Cell Electrofusion Chip Based On Micro-Cavity Microelectrodes

Hu, N., Yeungnam University
Qian, S., Old Dominion University
Joo, S. W., Yeungnam University
Originally presented on: 11/6/2013 13:58:00 - 14:20:00

A microfluidic chip with micro-cavity structures is designed and fabricated on a SOI substrate. More than 20,000 pairs of micro-cavity structures are patterned on the vertical sidewall of the serpentine-shaped microchannel. In each micro-cavity, the sidewall parallel to the microchannel serves as microelectrode; while the other two sidewalls, which are perpendicular to the microchannel, are fabricated by SiO_2 insulator. Such design is to focus electric field at the opening of each micro-cavity, where the cell-cell contact point and electrofusion occurs. Concentrated electric field at the contact point improves electrofusion efficiency and reduces the probability of multi-cell electrofusion. In addition, it reduces the imposed voltage for cell electrofusion and improves the viability of the fused cells. NIH3T3 and myoblast cells are used to test the microfluidic device. Under the positive dielectrophoretic force induced by AC signal (Frequency: 1MHz, Vp~p: 2V), almost all cells are captured into the micro-cavity structures and achieve 2-cell pairing with about 60% efficiency. Subsequently, DC electric pulses (9V, 50us) are applied to induce electrofusion. About 50% cells loaded into the microfluidic chip are fused into hybrid cells. Multi-cell electrofusion is not observed in the device. The cell pairing and fusion efficiency are higher than those obtained in traditional electrofusion methods and most existing microfluidic devices.

(482f) Dielectrophoretic Separation of Electroporated Cells

Cemazar, J., Virginia Tech
Kotnik, T.,
Originally presented on: 11/6/2013 14:20:00 - 14:42:00

An exposure of a cell to a sufficiently high external electric field results in electroporation, which alters electrical properties of cells and their dielectrophoretic response. We designed and fabricated a dielectrophoretic field-flow fractionation (DEP-FFF) device which allows for fractionation of irreversibly electroporated, reversibly electroporated, and nonelectroporated cells.

The chamber was constructed from a single Pyrex wafer with interdigitated Au electrodes, a spacer, and a top cover glass, making the events in the chamber observable under most optical microscopes. The dimensions were optimized based on numerical computations of the electric field, its gradient and the fluid-flow velocity profile. The electrodes were fabricated using photolithography. A double-sided self-adhesive tape of 100 μm thickness was used as a spacer, with an opening cut in its middle to form a channel, and with water-resistant acrylic glue of the tape holding the glass plates together and providing a tight seal. The glue loses its adhesive properties above 70°C, allowing for easy disassembly of the chamber in hot water and its thorough cleaning.

We first optimized the flow rate and electric field parameters for efficient DEP-FFF separation of moderately heat-treated CHO cells (50°C for 15 min) from untreated ones, with the former used as a uniform and stable model of electroporated cells. We then used CHO cells exposed to electric field pulses with amplitudes from 1200 to 2800 V/cm, yielding six groups containing various fractions of nonporated, reversibly porated, and irreversibly porated cells, and we tested their fractionation in the chamber. DEP-FFF at 65 kHz resulted in distinctive flow rates for nonporated and each of the porated cell groups. At lower frequencies the efficiency of fractionation deteriorated, while at higher frequencies the separation of individual elution profiles was further improved, but at the cost of cell flow rate slowdown in all the groups, implying undesired transition from negative into positive DEP, where the cells are pulled towards the electrodes at the bottom of the chamber. We tested DEP-FFF chamber with another cell line B16F1 and verified that our system can be used for different cell types. Throughput of this chamber is high, 10^5 cells can be separated in one batch in 6 min. Our results demonstrate that fractionation of irreversibly electroporated, reversibly electroporated, and nonelectroporated cells is feasible at a properly selected frequency.

(482g) Dielectrophoretic Detection Of Human Oral Cancer Using 3D Well Electrodes

Graham, K., University of Surrey
Mulhall, H., University of Surrey
Fedele, S., University of Surrey
Hughes, M., University of Surrey
Porter, S.,
Hoettges, K.,
Lewis, M., University of Surrey
Mcaul, J., University of Surrey
Kalavrezos, N., University of Surrey
Labeed, F.,
Originally presented on: 11/6/2013 14:42:00 - 15:04:00

Dielectrophoresis has for many decades been used in the discrimination between cancer and normal cells, commonly in the pursuit of the detection of tumour cells in circulating blood. However, these have yet to enter widespread clinical trials, largely due to issues of throughput, sensitivity and the number of cells required for analysis. Here we describe an alternative approach to the use of dielectrophoresis in performing cancer diagnosis, by distinguishing between oral cancer cells and normal oral epithelial cells from non-invasive oral brush samples taken from patients attending Head and Neck cancer clinics. The ready availability of tumour cells in high concentrations means that the device can be used to analyse the properties of the whole sample, eliminating the need for a separation stage. Using a 3D dielectrophoretic "well" chip, samples can be analysed with minimal preparation; discrimination is made on the basis of dispersion frequency analysis rather than interpreted dielectric data, eliminating the need for complicating factors such as measurement of radius. Our studies show that the test offers sensitivity and specificity equal or superior to current oral cancer diagnostics at substantially lower cost, with total consumables for sample testing costing $10 or less per test. The system also offers the added benefit of convenience - samples remain viable for up to five days after collection, allowing centralised testing and eliminating the need for each physician to have their own analysis device.

(527a) Big Advantages of Thinking Small
Muller, R.,
Originally presented on: 11/6/2013 15:15:00 - 15:40:00

It is now well past a half century since Richard Feynman presented his often-cited seminar bearing the provocative title, “There’s plenty of room at the bottom.” Feynman’s observation about the great engineering opportunity of the small scale was made several years before the invention of that great enabler, the integrated circuit, and more than a decade in advance of the first microelectromechanical systems. Both ICs and MEMS proved that indeed there is valuable room at the bottom, room for engineering systems that have provided legs for the profound upheaval of society itself through the internet, wireless communications, and related ground-breaking cousins. Today the design frontier is described in nano-scale units-- giving rise to application horizons that are expanding at ever-increasing rates. A major challenge for engineers today is to bring ‘up from the bottom’ the newly developed capabilities and to apply them in the solution of human-scale problems.

2013 Annual Meeting of the American Electrophoresis Society (AES)

(527d) Electroporation for Extraction of Intracellular Proteins and Genes

Lu, C., Virginia Tech

Originally presented on: 11/6/2013 16:30:00 - 16:55:00

Extraction of intracellular proteins and genes from cells is often an important first step for conducting molecular biology and proteomics studies. Electroporation is a simple and effective approach for these applications. Being a physical method, electroporation eliminates the use of chemical and biological lysis agents which require removal and may interfere with subsequent analysis. We demonstrated that electroporation was capable of releasing intracellular proteins from adherent Chinese hamster ovary cells while preserving the cell viability. By tuning the duration and intensity of an electric pulse, we were able to control the average amount of protein release and the percentage of viable cells after the operation. Furthermore, we showed that electrical lysis offered high efficiency for DNA extraction from both eukaryotic cells and bacterial cells that was comparable to the widely-used chemical lysis.

(527e) A Microfluidic Toolbox To Experimentally Probe Macromolecular Transport During Gel Electrophoresis: Insights and Opportunities

Ugaz, V. M., Texas A&M University

Originally presented on: 11/6/2013 16:55:00 - 17:20:00

This presentation describes our efforts harness miniaturization as a tool to obtain an improved fundamental understanding of biomolecular separations using gel electrophoresis. Our experimental approach is based on a microdevice design that incorporates on-chip arrays of electrodes, heaters, and temperature sensors. By embedding electrodes directly within the microchannel network and using a photocurable crosslinked polyacrylamide matrix, we are able to perform rapid high-resolution analysis of mobility and dispersion (band broadening) during separations of both single- and double-stranded DNA. These individually addressable electrode arrays can also be used to sequentially concentrate, focus, meter, and inject DNA into the separation matrix. In parallel with this experimental capability, we have also worked to obtain a detailed picture of average pore size and degree of polydispersity in crosslinked polyacrylamide hydrogel matrices. Our approach uniquely combines complementary information from transmission electron microscopy, in-situ dynamic mechanical analysis, and thermoporometry in a way that enables both the mean pore size and the pore size distribution to be quantitatively characterized. Insights obtained from these studies have enabled us to determine how to access a transport regime dominated by entropic trapping. Macromolecular transport in this regime can be synchronized to establish a resonance effect whose size-dependence can be exploited both to achieve improved separation resolution and as a new sensitive structural probe of DNA binding interactions.

(527f) Reminiscences About Electrophoresis

Garfin, D. E., American Electrophoresis Society

Originally presented on: 11/6/2013 17:20:00 - 17:45:00

It is an honor to receive the 2013 AES Service Award. I enjoyed my years on the AES Council especially those when I served as the society's president. Since I presented a short history of AES a few years ago, and since I was using electrophoresis even before AES was formed in 1979, and before my association with AES, I'd like to present a historical summary of some of the early electrophoresis techniques I learned in both research and production situations. The decades of developments I've witnessed and especially the current trend toward miniaturization attest to the vibrancy and richness of the field of electrophoresis. Beginning in the late 1960s I used paper electrophoresis to sequence oligonucleotides. Polyacrylamide gel electrophoresis (PAGE) and isoelectric focusing (IEF) emerged in the 1960s. In the 1980s western blotting, capillary electrophoresis, preparative PAGE, preparative IEF, and pulse field electrophoresis were developed. In the 1990s the precast gel market grew, as did the use of 2D-PAGE for proteomics. Many methods have faded from view, but others are still around. The field continues to evolve and I look forward to the emergence and growth of new and better procedures.

AUTHOR INDEX

AUTHOR INDEX

AUTHOR INDEX